INTEGRATED ENVIRONMENTAL MODELLING: DESIGN AND TOOLS

STUDIES IN OPERATIONAL REGIONAL SCIENCE

Folmer, H., Regional Economic Policy. 1986. ISBN 90-247-3308-1.

Brouwer, F., Integrated Environmental Modelling: Design and Tools. 1987. ISBN 90-247-3519-X.

Integrated Environmental Modelling: Design and Tools

by

Floor Brouwer

International Institute for Applied Systems Analysis
Schloss Laxenburg, Austria

1987 **KLUWER ACADEMIC PUBLISHERS**
a member of the KLUWER ACADEMIC PUBLISHERS GROUP
DORDRECHT / BOSTON / LANCASTER

Distributors

for the United States and Canada: Kluwer Academic Publishers, P.O. Box 358, Accord Station, Hingham, MA 02018-0358, USA
for the UK and Ireland: Kluwer Academic Publishers, MTP Press Limited, Falcon House, Queen Square, Lancaster LA1 1RN, UK
for all other countries: Kluwer Academic Publishers Group, Distribution Center, P.O. Box 322, 3300 AH Dordrecht, The Netherlands

Library of Congress Cataloging in Publication Data

```
Brouwer, Floor.
   Integrated environmental modelling.

   (Studies in operational regional science ; 2)
   Includes index.
   1. Regional planning--Environmental aspects--
Mathematical models.  2. Regional planning--Environmental
aspects--Netherlands--Biesbosch--Mathematical models.
3. Environmental impact analysis--Mathematical models.
4. Environmental impact analysis--Netherlands--Biesbosch--
mathematical models.  I. Title.  II. Series.
HT391.B763   1987      333.7'1'0724      87-3447
```

ISBN-13: 978-94-010-8117-7   e-ISBN-13: 978-94-009-3613-3
DOI: 10.1007/978-94-009-3613-3

PREFACE

In the mid 1980's - while a student at the department of econometrics at the Free University - I became an assistant at the Institute for Environmental Studies (IvM) of this university. My main task was assisting with the computational aspects of the project 'an integrated environmental model: a case study in the Markiezaat area'. A number of methodological problems were formulated during the operationalization phase of that project, such as the need to develop systematically an integrated model design and to look for means of handling different sources of information.

Prof. Dr. P. Nijkamp of the Department of Regional Economics and Drs. L. Hordijk - at that time leader of the economic-technological research group at the IvM - therefore initiated a project proposal to be supported by the Netherlands Organisation for the Advancement of Pure Research (ZWO).

Meanwhile I became an assistant to Prof. Dr. P. Nijkamp, surveying qualitative statistical developments in the field of regional inequality analysis. This inventory has been shown to be a relevant basis for the preparation of this book.

In spring 1982 I began working at IvM on the above project on integrated environmental modelling.

I am deeply indebted to Prof. Dr. P. Nijkamp for his support, enthusiasm as well as his criticism, during the whole period of this study. I would also like to thank Drs. L. Hordijk, who initiated the project, and his successor, Dr. W.H. Hafkamp. The grants from ZWO under project numbers 46-079 and 46-125 are also acknowledged, and greatly appreciated.

Most of Chapter 2 was prepared during a stay at the International Institute for Applied Systems Analysis (IIASA) in Austria. I would like to express my thanks to the Foundation IIASA-Nederland for their financial support during my three months at IIASA.

The numerical results in Chapter 7 are based on data from a parallel project at IvM on outdoor recreation and natural environment in the Biesbosch area. I gratefully acknowledge the coordinator of that project, Drs. S.W.F. van der Ploeg, for provision of this data.

Secretarial assistance at the Institute gave me the opportunity to finish this study in time, and I want to thank all of the secretariat for their willingness and patience to take care of and revise the various draft chapters of this study on a word processor.

Ms. A. Gilbert took care of the English style, and her suggestions have led to an improvement of the text. Her support is also greatly appreciated.

Last but not least I want to thank my wife and best friend Cobi for her moral and technical support as well as her criticism to improve the style of this book.

Floor Brouwer

Südstadt, January 1987

TABLE OF CONTENTS

VIII

CHAPTER 1. INTRODUCTION

1.1. REGIONAL ECONOMIC MODELLING

Macro-economic-oriented models have been developed in the past thirty years for many countries, both for free market enterprise economies as well as for centrally planned economies. Such models often include policy simulations on the basis of estimates of economic growth indicators such as national consumption, production, employment, etc. Meanwhile many sophisticated tools have been developed for economic models at the national level in order to incorporate, among others, dynamics, (dis)equilibrium, and technological development.

The main focal point of economic modelling approaches at the regional level is that such approaches aim at including the spatial aspects (i.e. the spatial distribution and interaction) of economic development. Regional economic models and sector models have increasingly become operational since the end of the sixties and the beginning of the seventies, as tools for regional planning and policy making, in order to investigate the consequences of alternative policy instruments or conflicting policy options. Such regional models may provide a framework for measuring the efficiency and equity of alternative economic policies. An example of a regional economic model developed at the end of the sixties in the United States can be found in Klein (1969), who specified a regional industry model which is linked to existing national economic models. Major research efforts concerning the development and application of regional economic models in that period were made in the field of models for land use planning and transportation analysis.

Generally speaking, the regional models developed in the late sixties and early seventies were analogous to and consistent with the existing national economic models, and hence were focusing attention on the spatial/temporal evolution of variables such as production, consumption, employment and income. During the first stage, the spatial dimension of the regional models covered primarily only one region.

After several years, the progress and increased efforts of research activities in the field of regional economic modelling led to the inclusion of multiregional phenomena and interregional linkages. Interregional economic models are models which include the linkages between regions in terms of, for example, interregional trade of commodities, flows of workers, or migration flows (see also Nijkamp et al., 1986 for a discussion of the main progress during the eighties and the state-of-the-art in the field of regional and multiregional economic models, and Courbis, 1979, for an application of an interdependent multiregional-national model with 22 regions for support in French planning).

Regional economic models permit various types of economic and demographic oriented activities to be taken into account (e.g., production, demand for production factors, balance between supply and demand of labour). Figure 1.1 shows a simple illustration of the scope of regional models including economic and demographic phenomena. A distinction has been made in this figure between a demand- and a supply-oriented economic component, and a demographic component.

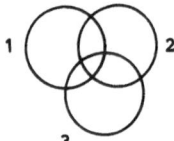

1 = demand oriented part (e.g., demand for land, capital or labour, and final demand).
2 = supply oriented part (e.g., supply of labour, capital, transportation facilities, energy,production).
3 = demographic part (e.g., population size).

Figure 1.1. Venn diagram of interrelations exhibiting the scope of a regional economic model.

A regional economic model may emphasize one or more of the three regional compartments mentioned in the figure. Intersections of the circles are achieved when at least two compartments are included in a model. A mixed supply/demand-oriented regional economic model for example, will emanate in the intersection area from 1 and 2 in the figure, without having any intersection with the demographic part. However, a regional economic model with a supply/demand-oriented component, while also including demographic phenomena, will be represented in the figure by means of the intersection of all parts. For an example of a supply/demand-oriented interregional policy model for the Netherlands see Lesuis et al. (1980), or for a demand/demographic-oriented regional model of regional labour markets in Austria, see Baumann and Schubert (1980). An example of a supply/demand- oriented model which also includes demographic variables can be found in Courbis (1979). A comprehensive international review of multiregional economic models can be found in Issaev et al. (1982).

The modelling approaches mentioned above may be used either for analytical purposes, or for policy and planning purposes. They only include a limited number of relevant economic phenomena. Limitations like scarcity of land, availability of energy and natural (renewable or nonrenewable) resources, environmental constraints, and quality of life were often not explicitly taken into account. Only since the end of the seventies such more comprehensive regional models have come into being.

1.2. ENVIRONMENTAL ISSUES IN REGIONAL ECONOMIC MODELLING

The major focus of economics is upon production, choice and distribution of goods and services. In particular it concentrates on the manner in which man employs scarce resources in alternative ways.

Concern for environmental decay in our society already dates back to at least the thirteenth century when a city, say London, suffered air pollution problems from the burning of soft coal. The interdependence between economic activities and environmental phenomena might be seen also at least one hundred years ago when the increased demand for wood and agricultural products in European countries have led to environmental impacts such as deforestation in Far East countries (Clark, 1986). However, the inclusion of environmental phenomena in economic analysis dates back to the nineteenth century when Malthus and Ricardo described a scenario of limitations on the growth of population, and the availability of agricultural land. Malthus defined human well-being in terms of the ratio of environmental resources (such as food production) to human population.

The major environmental issue since the middle of the twentieth century is pollution (either of air, water, or land), because that period was characterized by rapid economic growth together with a growing concern for environmental decline and for limits of natural resources. However, it was not before 1960 that 'pollution was widely recognized by economists and others as a serious threat to human well-being' (Fisher and Peterson, 1976, p. 3). Since the end of the sixties environmental problems have come to the fore as a major issue in our society. This has also led to the emergence and popularity of environmental sciences.

Environmental sciences are typically multidimensional because the pertaining phenomena emerge from different disciplines, such as economics, ecology, demography, geography, natural science and socio-political science, and also cross the boundary of single disciplines (see also Nijkamp, 1977). Urban quality of life for example, is based upon a multidimensional environmental phenomenon that can only be adequately represented by means of a multidimensional vector profile with elements containing the quality and size of dwellings, the availability of natural parks and recreational facilities, transportation and educational facilities, etc. The environmental phenomena are therefore pluriform in nature and emerge from various disciplines.

One of the first attempts to include environmental issues in the frame of (regional) economic modelling is the input-output (or I/O) approach developed by Isard, Leontief and Cumberland (see for example Isard, 1968). The input-output models provide a description of the production side of an economy, with sectoral linkages between input requirements and deliveries of output. Emissions of pollutants are incorporated in an I/O model, and are linked to the economic production sectors.

The environmental issues in regional economic modelling arise in a new branch of economics, viz. environmental economics. This (sub)discipline mainly focuses on the economic aspects of pollution and of environmental quality. Pollution of air, water and soil are the major environmental impacts of human economic activities such as production and consumption. Relevant environmental resources to be included in the framework of environmental economics are land, soil, water, air, plant and animal life (Lakshmanan and Bolton, 1986).

An integrated modelling approach for the inclusion of environmental issues in regional policy impact analysis is a new branch in the field of environmental economic research, initiated mainly in the mid seventies.

1.3. INTEGRATED ENVIRONMENTAL MODELLING

Since the early seventies a great many efforts have been made at modelling the interactions between economic activities, demographic development, and ecological processes at a regional scale. The need to include the spatial aspects and the spatial distribution led to the development of such models at the regional level (see also the progress in the field of regional economic modelling as mentioned before in Section 1.1). The main reason for modelling the complex connections between economy and ecology was the growing awareness of the far reaching environmental impacts of economic activities. The formal representation of such modelling approaches will be coined here integrated environmental models (shortly, IEMs).

Environmental phenomena such as emissions to the air, depend on spatial variation, because of relationships with industrial activities and of population density.

The increased interest in this research area is inter alia reflected in the UNESCO programme Man and Biosphere, which was initiated in 1971, and is aimed to 'build a research bridge' between ecological, economic, and social sciences as well as to elaborate a methodology for linking ecological processes and (socio)economic phenomena with each other.

Many attempts have been made in the seventies to model the interactions between economic and environmental processes at the national level. An example of a large-scale national-regional integrated modelling approach is termed the Strategic Environmental Assessment System (SEAS) model. This model has been developed for the Environmental Protection Agency in the United States, and it represents a set of interlinked economic, environmental and energy models. The regional impacts are also included in the SEAS model. The SEAS model is based on a dynamic input-output model of the United States with 185 sectors that release pollutants into air, water and land. It has been used to evaluate the impacts of federal environmental and energy laws and policies (Lakshmanan and Bolton, 1986; Lakshmanan and Ratick, 1980).

A multiregional model for the economic, environmental, and energy consequences of national and regional policies is developed by Lakshmanan (1983) at the US Bureau of Economic Analysis. The model, called a Multi Regional Model of the Economy, Environment and Energy Demand (MREEED), is a tool for the assessment of economic, environmental, and energy impacts of national and regional policies.

In the sequel to this study an IEM is defined as a model which represents the regional structure of economic, ecological, and socio-geographical processes as well as their mutual relationships (see also Nijkamp and Opschoor, 1977). Such a model may be used for description, evaluation, or forecasting.

1.4. OUTLINE OF THE STUDY

The aim of this study is (i) to develop some new methods for designing an IEM, (ii) to develop some mathematical tools for the operationalization of an IEM, and (iii) to develop an integrated environmental modelling approach for the Biesbosch area in the Netherlands. In order to do so, this study has been subdivided into three main blocks. Table 1.1 shows the outline of the study. The three main parts, each consisting of two chapters, will be discussed briefly below.

Part A is an <u>introduction to integrated environmental modelling</u>. A survey of fourteen IEMs will be discussed in Chapter 2, which cover various environmental categories related to water, air, and land use. This survey representing the state-of-the-art, also demonstrates that all types of environmental categories can, at least in principle, be implemented and operationalized by means of IEMs.

The main characteristics and the correspondence between the models out of the survey will be evaluated in Chapter 3 with respect to nine features and criteria characterizing the IEMs.

Part A shows that a systematic approach is necessary to operationalize the conflicting issues included in the framework of an IEM.

Part B deals with <u>the methodology of integration and mathematical tools</u> to operationalize such IEMs. Systems theory, which will be discussed in Chapter

Table 1.1. Outline of the study.

Chapter 1 : introduction
Part A : introduction to integrated environmental modelling approaches: a survey with an evaluation
Chapter 2 : a survey of IEMs Chapter 3 : an evaluation of the survey
Part B : methodology and mathematical tools to operationalize an IEM
Chapter 4 : a systems approach to an IEM Chapter 5 : statistical and econometric tools to operationalize an IEM
Part C : an integrated environmental modelling approach
Chapter 6 : design of an IEM; causality/qualitative analysis of an IEM Chapter 7 : outdoor recreation in the Biesbosch area
Chapter 8 : conclusions

4, is a helpful methodological framework to integrate phenomena originating from various disciplines. A recently developed framework of integration, called the satellite design, will also be presented in that chapter. This satellite design of an IEM is an hierarchical model structure with systems and interrelated sub-systems, which is based upon three levels, viz.:

(i) the core level, which is considered to be the central part of the analysis. The determination of the core is based on a priori knowledge, or a priori assumptions concerning impacts;

(ii) the second level, which represents the aspects from the core level, related to the other phenomena in the analysis;

(iii) the third level, which consists of the relationships between the phenomena that are not covered by the core level.

Some statistical and mathematical tools to operationalize an IEM will be presented in Chapter 5, dealing with various levels of information, viz.

(a) the available information in IEMs may be not known with sufficient precision to draw reliable quantitative (cardinal) conclusions concerning analytical or policy aspects. Graph theory may then be a useful tool in case of:

  - binary information on the causal structure to establish an hierarchical ordering of stimulus-response patterns;

  - qualitative information on the stimulus-response patterns to solve a set of equations and to determine the qualitative impacts of policy variables. This is also be called the sign-solvability approach.

(b) the different data sources of IEMs may provide information measured at

various metric scales, varying from a nominal, an ordinal to a cardinal
scale of measurement. A family of interrelated statistical models is
covered by the generalized linear model (or GLM) approach, including a
wide range of linear models for metric and non-metric information.

(c)   the information on economic, social, and environmental phenomena may be
       multivariate and complexly structured in nature. Scaling methods are
       often used to extract and quantify the relationships among the multi-
       variate items.

The design and mathematical tools with regard to an <u>integrated environmental
modelling approach</u> for recreational activities in the Biesbosch area in the
Netherlands will be presented in Part C of this study. The design of an IEM
will be defined in Chapter 6 for the Biesbosch area by making use of the
satellite concept of integration. The causal model structure of the stimulus-
response patterns will also be discussed by making use of graph theory (in
terms of directed graphs) to trace out the hierarchical nature of variables.
Information, in addition to the binary relationships on the causal model
structure, may also be available with respect to the signs of the impacts
between variables. The sign-solvability approach (interpreted now in graph
terms by means of signed directed graphs) will be used for a simulation model
to determine the qualitative impacts of stimulus patterns on the response
variables.
Chapter 7 shows some mathematical tools which are used in case of modelling
environmental phenomena in the Biesbosch area. Special emphasis will be
placed on a spatial characterization of the various aspects related to outdoor recreation, viz.:

(a)   an exploratory analysis of the recreational activities, with respect
       to, among others, the type of boats used by recreationers, and the
       distance to the home address. This approach makes use of log-linear
       models;

(b)   an explanatory analysis with respect to the place where the daily pur-
       chases will be bought, in terms of the number of days of staying in the
       area and the distance to the home address. This approach makes use of
       linear logit models;

(c)   a multivariate analysis with respect to the relevance of various pheno-
       mena to spend a period in the Biesbosch area, such as environmental
       phenomena, and phenomena regarding to the availability of recreational
       facilities. The approaches make use of a HOMALS procedure and a multi-
       dimensional scaling procedure;

(d)   a stimulus-response analysis with respect to the type of boats, and the
       various types of recreational activities in the Biesbosch, which makes

use of a path analysis model.

This study will be evaluated and finished with Chapter 8 where some concluding remarks will be presented concerning the design of an IEM and the tools to operationalize an IEM.

PART A:

INTRODUCTION TO INTEGRATED ENVIRONMENTAL MODELLING:
A SURVEY AND AN EVALUATION

CHAPTER 2. INTEGRATED ENVIRONMENTAL MODELS: A SURVEY

2.1. INTRODUCTION

The main aim of treating these models here is to indicate the wide range of situations in which integrated environmental models (abbreviated as IEMs) have been applied in the past ten years (e.g., in water resource management models, land use planning models, agricultural, regional and urban planning, as well as ecosystem management planning models). Fourteen integrated environmental models will be discussed. This is a small but fairly representative sample of models which have actually been operationalized. The next chapter evaluates the strong and weak points of each of these models. The models originate from various countries with different fields of application, and use various mathematical tools for integrating the successive modules. Examples of such mathematical tools are, among others, optimization, simulation and evaluation techniques. An exhaustive survey of IEMs is given by Braat and van Lierop (1984). Such models are characterized by a multidisciplinary approach to environmentally relevant phenomena originating from different disciplines. There are links - either direct and/or indirect - between various constituent modules which have their background in different disciplines such as (regional) economics, ecology, demography, recreation, geography etc. In the sequel of this chapter, a module will be interpreted as a set of interrelated - often monodisciplinary oriented - variables which have their background in a specific, identifiable part of a compound environmental phenomenon. Brief background information on the origin of the 14 models to be discussed here is given in Table 2.1.

Table 2.1. Background information on 14 integrated environmental models.

Section	Title of study/paper	Authors	Country
(2.2)	Multiobjective control of nutrient loadings into a lake	L. Duckstein, I.Bogardi, L. David	Hungary
(2.3)	Simulation of regional development and fishery activities	W.R. Boynton, D.E. Hawkins, C. Gray	U.S.A.
(2.4)	Policy analysis of water management	B.F. Goeller et al.	Netherlands

Table 2.1 (continued).

Section	Title of study/paper	Authors	Country
(2.5)	Policy analysis of the Ooster-schelde	B.F. Goeller et al.	Netherlands
(2.6)	Impact analysis of flood plain models	K.C. Tai	U.S.A.
(2.7)	An economic-ecological model for land-marine integrated develop-ment	S. Ikeda, H. Hakanashi, Y. Nishikawa	Japan
(2.8)	Environmental ecosystem quality management model	R.A. Kelly, C.S. Russell W.O. Spofford	U.S.A.
(2.9)	TLM of a national-regional economic-environmental model	W. Hafkamp	Netherlands
(2.10)	Systems approach of economic-environmental-energy analysis	I.M. Andréassen, H. Ahlblom, A-M. Jansson, G. Spiller, J. Zucchetto	Sweden
(2.11)	Simulation of impacts of herbicides on the environment and agriculture	M.T. Brown	Vietnam
(2.12)	Economic-ecological analysis by simulation and optimization	S.C. Lonergan	Canada
(2.13)	Urbanization and environmental planning and design	P. Rogers, C. Steinitz	U.S.A.
(2.14)	Integrated regional environmental model for land use planning	J.W. Arntzen, L.C. Braat	Netherlands
(2.15)	Interactions between economy, ecology in a region of intensive agriculture	N. Müller, B. Thober, H. Lieth, S. Fabrewitz	F.R.G.

The models in Table 2.1 have been classified according to their main environ-mental phenomena, viz. water-related, air-related or land-use related envi-ronmental phenomena.

The integrated models which will be discussed in Sections 2.2 to 2.7 deal with environmental aspects related to water (e.g., water quality, water shortage or water flood). The models to be discussed in Sections 2.8 and 2.9

deal with environmental aspects related to air (e.g., residuals, emission
and dispersion) and the models which will be discussed in Sections 2.10 to
2.15 are related to various types of land use (e.g., agricultural activities,
regional physical planning).

Two general characteristics of the 14 selected IEMs can now already be mentioned on the basis of the information contained in Table 2.1. First, all of
them have direct or indirect feedbacks between an economic and an ecological
module. The IEMs may be termed also economic-ecological models because of the
emphasis which is placed upon the integrated analysis of relevant phenomena
from both disciplines.

Second, the models are operationalized with impacts measured at a regional
level. However, it should be noted that the size of a region depends on the
country concerned; in the U.S.A., for example, the size of a state may be
equal to a small European country. The size of a region in the above mentioned sample is relatively vague, as it varies between the national level and
the local level; no further distinction can be made to characterize a region
precisely. Some models have also links between regions and are therefore
called multiregional models (viz., the models in Sections (2.4) and (2.9)).

The main points of reference for discussing each IEM from Table 2.1 are:

(1) the type of modules that have been applied when operationalizing an inte-
    grated environmental model. One may consider modules representing pheno-
    mena or processes from various disciplines, such as (spatial) economics,
    ecology and demography as well as transportation analysis and recreation
    analysis.

(2) the characterization of the modules, especially features of the links
    between the modules. A distinction can be made between interdisciplinary
    relationships (links between variables which cross the boundaries of
    different modules) and intradisciplinary relationships (links between
    variables from one module).

(3) the time horizon (also called temporal coverage) of the model analysis as
    well as the time dimensions (or temporal resolution) from the separate
    modules;

(4) the spatial dimension of the modules, subdivided into spatial coverage
    and spatial resolution;

(5) the dimensions of the variables to operationalize an IEM. The variables
    can be characterized in, for example, monetary terms, energy flows, nom-
    inal terms or dimensionless figures. The main question to be discussed
    here is whether the variables are denoted by their natural dimensions
    which originally characterize these variables or whether they are trans-

formed into other dimensions. A transformation of different dimensions
into equivalent dimensions may be useful in case of linking such variables.

(6) the use of **mathematical tools**. Examples of modelling approaches are optimization (e.g., linear or nonlinear mathematical programming), simulation
(e.g., differential or difference equation models),input-output analysis,
statistical or econometric modelling, scenario analysis, or decision
modelling (e.g. multiobjective decision analysis).

With these points of reference in mind, a precise presentation of 14 integrated environmental models is given in the next sections.

2.2. MULTIOBJECTIVE CONTROL OF NUTRIENT LOADINGS INTO A LAKE

The study was initiated to analyse the possibilities for control of the level
of nutrient loading in water resulting from agricultural activities and regional development (see Duckstein et al., 1980, 1982). The control of nutrient loadings has a tradeoff relationship with agricultural benefits and
regional development. The aims of the analysis were to obtain an analytical
framework for reducing phosphorus loadings from a watershed to a waterbody,
and also to control nutrient loading in water. The Lake Balaton area in Hungary was used as a case study.

In the analysis it is shown that an increase of phosphorus loading is determined by, among others, commercial and natural fertilizers, and has a negative effect on water quality because of eutrophication (*). The two basic
elements of the analysis (economic growth and reduction of phosphorus in
water) are mutually dependent, and also are conflicting in nature since economic growth (denoted by an increase of agricultural benefits) may require
increased use of fertilizers. A reduction of phosphorus loading may be achieved when a higher level of water quality control is introduced, which will
reduce economic benefits through rise of the costs of agricultural activities. The mutual dependence between the reduction of phosphorus loading and
the benefits of agricultural activities is analysed by means of a multiobjective framework. An ideal solution can only be obtained when both ojectives
are optimized independently from each other. However, such a solution is
nonfeasible because of the dependence between both objectives. For this reason, a compromise programming solution has to be determined, which can be
obtained by minimizing the distance between the ideal solution and the set of
feasible solutions. The aim is to achieve an 'acceptable' level of both the
phosphorus loading in water and the economic benefits. The 'acceptable' level

(*) (a heavy growth of algae which especially occurs in still waters).

has to be chosen by policy makers. The compromise programming approach with two objectives makes use of the following objective function:

$$\text{Minimize} \quad L_p = \left(\sum_{i=1}^{2} \alpha_i^p \frac{|Max(i) - f(i)|^p}{|Max(i) - Min(i)|^p} \right)^{1/p}, \quad p \geq 1 \qquad (2.1)$$

where $f(i)$ is a feasible solution of objective i, $Max(i)$ and $Min(i)$ are the maximum and minimum values of objective i ($i = 1,2$), α_i the preference weight or importance of the objective and p a parameter of the preference weight as well as of the Minkowski distance metric for both the numerator and the denominator ($p \geq 1$). The preference weights are normalized by $\sum_{i=1}^{2} \alpha_i = 1$ and denote the relative importance of each objective. The parameter p has been put equal to one by the authors; this means that the numerator as well as the denominator are both absolute values of differences.

The two parts of formula (2.1) represent numerical operations on variables in order to transform them into dimensionless figures. The natural dimension of the reduction of phosphorus loading from commercial and natural fertilizers used in agriculture is denoted in terms of percentages and can now via (2.1) be compared with the yields from agricultural activities whose natural dimensions are originally denoted in terms of money.

Other interrelated sectors of interest can be used analogously in the multi-objective approach by using a transformation into dimensionless figures. Examples are, among others, industrial development, drinking water supply or flood control.

The analysis for relating agricultural activities to lake eutrophication is summarized in Figure 2.1. An indirect relationship is also considered in this figure, viz., between agricultural activities and lake eutrophication. Lake eutrophication is the result of phosphorus loading and has been used here only as a background relationship because it is not included in the compromise programming procedure. It should also be noted that no temporal dimension is included here and that the compromise programming procedure is comparatively static in nature. The spatial coverage includes the Lake Balaton area in Hungary, without a further spatial subdivision being made within the area.

An advantage of the compromise programming procedures in this example is that the compromise solutions for alternatives whose variables originate from different disciplines (i.e., economics and ecology) are related to each other in a rather straightforward way. However, a disadvantage of this approach is that in the analysis no distinction can be made for the difference in qualitative nature of the variables as the units of measurement are directly transformed into dimensionless figures.

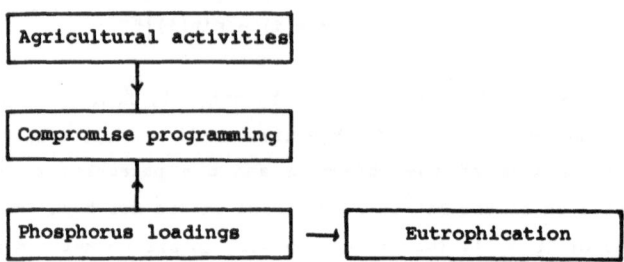

Figure 2.1.Trade off between agricultural activities and phosphorus loadings.

2.3. SIMULATION OF REGIONAL DEVELOPMENT AND FISHERY ACTIVITIES

The study area - the Franklin County in Florida (USA) - is a coastal region with a local economy which is traditionally dependent on water-resource-based activities, with emphasis on the oyster fishery activities in Apalachicola Bay (see also Boynton et al., 1977). The local oyster industry accounts for about 50% of total county income, and so is extremely important for the regional economic value added. The fishery industry is related also to environmental aspects, because the yields from such industrial activities depend heavily on water quality (e.g., organic matter concentration).

The aim of the study was to analyse whether additional economic activities could be made without getting into conflict with the fishery-based economy, which at worst might even eliminate the fishery economy in the Apalachicola Bay. For this reason, the analysis must evaluate the sensitivity of the local oyster fishery to the additionally planned developments in the region, and must incorporate the effects of changed levels of natural inputs and urban development. The change of urban development in the region mainly consists of recent and expected changes in tourism and the cattle industry, as well as additionally built retirement houses and expanded shipping facilities. The goal of the analysis is to obtain a balance between such economic activities and its ecological impacts.

An energy model has been developed for this region, which includes all relevant regional aspects, denoted in terms of energy, materials and money flows. The model has been operationalized at the county level, a more aggregated level than the local level at which the fishery activities take place. The reason to use a more aggregated level than the local level was that it might

be easier to include all relevant direct as well as indirect economic and
ecological consequences at such a higher scale level. The detailed interactions contained in the regional model are investigated by modules with a
lower scale.

A regional model of the Apalachicola Bay consists of three interrelated main
modules:

(i)   the Apalachicola Bay module with its biotic and physical parameters;

(ii)   the oyster industry module;

(iii) the county development module.

A small number of pathways of energy and money flows from the model structure
is illustrated in Figure 2.2.

county development module

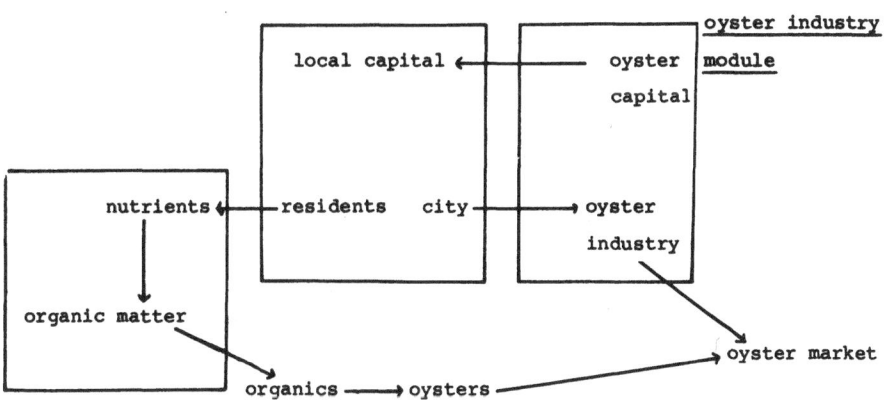

Figure 2.2. Illustration of the pathways between the three modules.

The intended regional economic development in the Franklin County (viz. changes in the cattle industry, tourism, construction of retirement homes and
expanded shipping facilities) is denoted by the county development module,
and it is at a more aggregated spatial level than the modules of the oyster
fishery activities. The ecological and economic framework of the Apalachicola
Bay is represented successively by means of modules (i) and (ii). The flows
between variables from the three modules is represented either in money
terms, in energy terms or in number of persons (Boynton et al., 1977, p.
484). The model was simulated by means of nonlinear differential equations.
See for example the equation which represents the rate of change of the oyster industry capital:

$$\dot{Q}_9 = \frac{dQ_9}{dt} = k_{27} \, Q_{10}Q_{11} - k_7 Q_9 - k_{28}Q_9, \qquad\qquad (2.2)$$

with: Q_9 = oyster industry capital;

           Q_{10} = oyster industry structure;

           Q_{11} = standing crop of oysters in the bay;

           k_7, k_{27}, k_{28} are parameters.

Decisions whether or not to include some variables in the model were based on literature surveys as well as fieldwork conducted in the bay. The model consists of the regional energy sources which are dominant and/or expected to change. Model results were tested against field observations with historical data to determine whether the selected factors are the relevant ones and whether the coefficients are correct. Boynton et al. (1977) only mention "that major interactions and forcing functions have been included" (Boynton et al. (1977), p. 493).

The model has been simulated for the period 1970-2000, and an extended model simulation with the year 2120 as time horizon has also been made (150 years forecasts). Simulation results have been used already in county planning decisions. However, it should be noted that all external forcing functions are assumed to be under constant levels, with 1970 as year of reference. The consequences of different management decisions have been explored by the model simulations; for example, a one year closure of the Bay with no oysters allowed to be harvested.

2.4. POLICY ANALYSIS OF WATER MANAGEMENT FOR THE NETHERLANDS

The PAWN-study concerning the water managemant analysis for the Netherlands was a joint research effort carried out by the Netherlands Rijkswaterstaat, the Delft Hydraulics Laboratory, and the Rand Corporation, Santa Monica,California (USA); this project started in 1977 and was finished in 1982.

The primary task of the project consisted of two parts (see also the summary report of the PAWN-study, Goeller et al., 1983):

    (i)   development of a methodology for assessing the multiple consequences of water management policies;

    (ii)   application of the methodology obtained from the project to develop alternative water management policies for the Netherlands and comparison of their consequences.

The methodology is shown in Figure 2.3 with the links given between the modules.

The water management problems of the Netherlands can be distinguished into shortage, salinity (salt water from the North Sea and the increasing salinity of the Rhine river are the main sources of salinity), other water quality issues and flood. Water salinity for example, can damage crops and has a negative influence on the natural environment and human health.

The above mentioned water oriented management problems are sometimes rather paradoxical, as some regions in the Netherlands suffer from a shortage and others from an excess of water. An increase in population and industrialization may cause a shortage of water, water quality problems and also environmental and distributional problems. All such events have consequences for both the regional and national level of water management (more details about regional and national water management policies can be found in Goeller et al., 1983).

The problem of shortage or flood of water in various regions shows an advantage of analysing water management policies at a regional level. Regional differences might be averaged out when the spatial level is the national level because of the aggregation process. The spatial coverage of the analysis includes the whole country and the spatial resolution is based on 77 regions.

The policy options contain a mixture of tactics, each of which may affect water management. Examples of such tactics are an investigation of new flushing rules for lakes, or an expansion of the water supply capacity in some areas. Such tactics influence demand and/or supply of water. A combination of several tactics will be called a strategy.

The analysis of water management does not consist of one large model. Each box in Figure 2.3 denotes a module which is linked directly to the water distribution module (DM) and has been analysed separately from the other modules. The methodological aspects of the PAWN-study are discussed in detail by Légrády (1983). The water distribution module is the core of the analysis because of its direct impacts with all other modules and because it simulates the distribution of water between the supply and demand part. A balance between supply and demand from the different sectors is obtained from and regulated by the DM. The DM gives information on flows, levels, discharges, pollutant concentrations, shipping depth, operating costs for sprankling and for tactics, low water shipping losses, shortages and salinity losses for agriculture. Figure 2.3 presents two circles representing supply and demand. The modules belong to one circle, depending on whether the module affects the supply of or the demand for water.

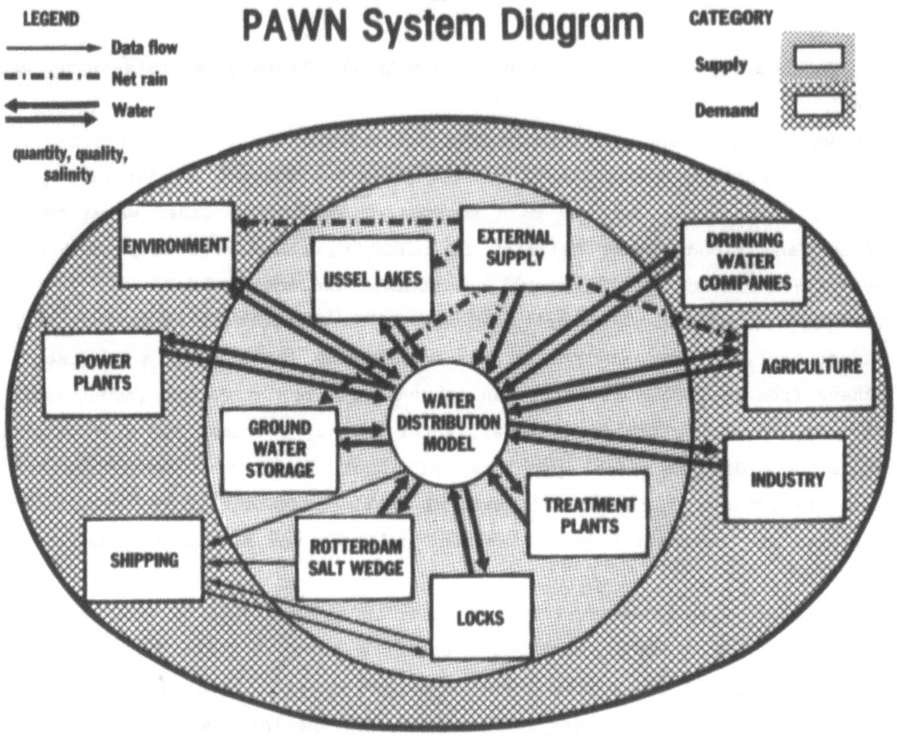

PAWN System Diagram

Figure 2.3. PAWN-system diagram.

(Source: Goeller et al., 1983, p. 57)

The demand for or supply of water is computed for each ten-day period (which is the temporal resolution of the analysis), and for all sectors. The temporal coverage is one year, since the DM run simulates the distribution of water for a calendar year (with 36 periods of ten days).

The DM consists of three main elements:

(i)   distribution of water in all 77 regions during ten days as a result of the supply and demand factors;

(ii)   calculation of monetary losses to agriculture and shipping because of a shortage of water or salinity. The losses are determined by relating the actual results during a decade concerning the distribution of water with the ideal situation in which neither of the users related to agriculture and shipping suffer from any losses due to water shortage and salinity;

(iii) determination of concentrations for six pollutants, viz. salts, heat, phosphate, biochemical oxygen demand (BOD), nitrogen and chromium. Salt concentrations are determined for each period of ten days; the other concentrations are obtained optionally in the analysis. The chloride concentration is used in the environmental module as a measure of salinity.

An environmental module is included; the environmental aspects of water management are dealt with by imposing water quality standards because "it is simply too difficult to treat the environment more directly" (Goeller et al., 1983, p. 157) due to a lack of reliable data. Determination of such water quality standards consists of two steps, viz.

(1) transportation of pollutants in the PAWN-network, as implemented in the DM-module;

(2) eutrophication in lakes which follows from a nutrient level, an algae bloom level (in terms of chlorophyll concentration) and a dissolved oxygen level.

Direct links exist only between the water distribution module and the other modules. The water distribution module acts as interference and is the key-factor for the other modules. No direct links exist between the other modules. The indirect links between the environment module and the industry module for water management policies, for example, can be explained in the following way. Groundwater is used by industries in various production processes as well as for cooling. The use of groundwater is regulated by the DM-module, and a direct link exists for that reason between the industry module and the DM-module. The environment module also has direct links with the DM-module because of the determination of the quality of groundwater. So the use of groundwater by industry is only linked indirectly to the environment module.

The monetary and non-monetary impacts of different water management policies are also investigated. The impact assessment analysis is a static and linear representation of a market economy with supply and demand functions. A water management policy is interpreted as a change in one or more supply functions.

In general, two characteristics of a detailed policy analysis for water management are:

(i) complexity of a system (like the PAWN-systems diagram) because of the numerous scenario alternatives;

(ii) uncertainty of the system because of incomplete knowledge.

Both characteristics depend on shortcomings in information because the data are either of poor quality or not available at all. Nonlinearities appear for the same reason in only one module in a policy design part. All other possible nonlinearities are linearized for computational reasons.

The methodology used may be rather simple, because it mainly deals with data processing. Much emphasis was placed upon the interaction between decision makers and policy analysts during the period of model development and operationalization.

2.5. POLICY ANALYSIS TO PROTECT AN ESTUARY FROM FLOODS

The RAND-Corporation, Santa Monica, California (USA) prepared a study of the Netherlands Oosterschelde, in collaboration with the Netherlands Rijkswaterstaat, to analyse the protection of an estuary from floods (see the summary report Goeller et al., 1977), called the POLANO-project. Three alternatives have been evaluated for the Oosterschelde region:

(i)   a closed case, with a dam across the mouth of the Oosterschelde, to close off the estuary from the sea. The main effect of this alternative might threaten the rich ecology as well as the oyster and mussel industry;

(ii)   a storm-surge barrier (SSB) in the mouth of the Oosterschelde, which may be closed during severe storms, and large gates which are open under normal weather conditions to allow a reduced tide to pass into the basin;

(iii) an open case, which leaves the mouth of the Oosterschelde open to maintain the tide. This alternative is based on the construction of a system of large dikes around the estuary to protect the land from floods.

There are many different consequences (usually called impacts) for each of the three mentioned alternatives. These may be subdivided into four groups, viz. the financial costs, security of people from flooding, ecology of the region, and the other economic and social impacts. These consequences are measured in different dimensions, while some cannot be quantified at all and can only be represented in terms of 'good', 'bad' and 'worse'. The impacts are represented therefore in different units (e.g., money versus number of species). Clearly, one of the three alternatives has to be selected by the policy makers. The various impacts of the alternatives are displayed on a score card that shows by colour code each alternative ranking for a particular impact, with impact values represented in their natural dimensions. An advantage of the scorecard approach is that it is possible to include a wide range of impacts while showing the strong and weak elements of various alternatives. The decisionmaker(s) can then add their value judgements about the relative importance of the different impacts, give weights to the impacts and select one of the alternatives.

The major difficulty which remains when the impacts of the alternatives have been assessed is to obtain a synthesis of the numerous and diverse impacts which are denoted in different units of measurement. Such a synthesis is necessary to be able to select one of the alternatives. Two selection procedures can be used:

(1) an <u>aggregate</u> approach, in which each impact is weighed by its relative importance and transformed into a single unit of measurement
such as money or utility. The alternatives are then compared with
each other on the basis of this aggregate measure. Some disadvantages
of this approach are:
- loss of information by the aggregation process. One alternative,
  for example, may lead to environmental problems, another one having
  high financial costs. Such information will be lost by an aggregation procedure.
- weights can implicitly be attached to the different impacts when
  impacts are aggregated. However, a single measure for an alternative depends strongly on such a weight set.
- conflicting goals may arise when discriminating between the alternatives, if there is more than one decisionmaker (as in the POLANO-
  project).
- the outcome of each impact is considered to be independent of the
  outcome of all other impacts.

(2) a <u>disaggregate</u> approach, with impacts expressed in natural units (in
monetary terms or in physical units); some impacts are quantitative
estimates and other impacts are denoted by qualitative comparisons.
The disaggregate approach with variables expressed in their natural units of
measurement is used in the POLANO-study. A disadvantage of this approach may
be the degree of detail which makes it difficult to identify patterns.
The four groups of impacts mentioned before will now briefly be discussed:

(1) <u>Financial costs of the alternatives</u> deal with the direct financial
costs necessary to construct and operate the alternative cases, viz.
the costs necessary to construct each of the three alternatives, the
annual costs to operate each of the alternatives after the construction period, and the maximum annual expenditures during the construction period. The information is denoted by the scorecard in monetary
terms.

(2) <u>Flood security impacts</u> or the regional security from floods. The
safety of each alternative is determined separately for two different
time periods, viz. the transition period (before construction is
completed) and the long run. Such impacts contain, for example, the
expected damage during the transition period emerging from land area
flooded.

(3) <u>Ecological impacts</u>, which concern the consequences for the biological
species in the Oosterschelde in response to each of the three alternative plans. Limitations in the scope of the ecological impacts
are:

(i)   geographical limitation, because the ecology of the salt-water area is the only one included, while the ultimate fresh-water lakes are excluded from the analysis.

(ii) biological limitation, because more than 2000 species are aggregated and clustered into 18 ecological groups.

(iii) temporal limitation, because only the long-term average abundances of the ecogroups are predicted. The daily and seasonal fluctuations are averaged out in this model.

The ecological impacts are estimated directly from sampled data. Tentative assumptions have been made concerning those impacts when no sampled data were available. One of the assumptions made, for example, deals with the species density in the Oosterschelde. This density is considered to be the same as in a similar estuary close to the study-area.

The ecological module has been calibrated and validated in the following way: to calibrate the module, the parameters and impacts between variables are reestimated a number of times by modification or by replacement of some of the assumptions. The model validation process determines how the model reflects reality. The model has been validated in this study by operationalization under comparable circumstances and the model has for that reason been simulated also for another estuary.

The implication and conclusion of the validation procedure is that:

(a)   predictions have a lot of uncertainties which reflect the uncertainty and lack of precision of the available data.

(b)   a distinction can only be made between two alternatives, denoted by the scorecard, when the results from each impact are very different.

(4) Additional impacts which are economic and social in nature (for example, commercial fishing, inland shipping and recreation). Recreational impacts were determined for each alternative by comparison of two alternative (with-without) policies, viz. one with a no-investment policy and one with an unrestricted investment policy. The information about the impacts from investment policies on recreational activities was obtained from field observations and personal communications with recreation authorities. A general problem concerning these impacts was the limited availability of reliable data as well as the uncertainty about the future because of lack of controlled experiments.

The selection of an alternative was obtained by weighing the impacts which denote the decisionmakers value judgements about the relative importance of the different impacts. Due to the complexity and uncertainty of the great many impacts in this policy analysis, tentative assumptions about the impacts between dependent variables have been made when no sampled data were available. The lack of information occurred especially in the ecological module.

2.6. ECONOMIC-ENVIRONMENTAL-ECOLOGICAL IMPACT ANALYSIS OF FLOOD PLAIN MODELS

The aim of flood plain management models is to evaluate the impacts of project alternatives on the economy, the environment and ecology (Tai, 1979). The economic impact of a selected project alternative is here evaluated when a mixture of flood control measures have been established (e.g., flood warning, flood plain zoning and insurance). The impact on the environmental quality of the main river has been evaluated in the Arkansas area after the economic impacts had been determined.
The different steps of the analysis are summarized in Figure 2.4.
No temporal dimensions are included in the input-output analysis. The input-output analysis assumes constant technical coefficients. The three steps of the flood plain management analysis, (see Figure 2.4) making use of an input-output approach, are analysed separately and will be discussed briefly below:

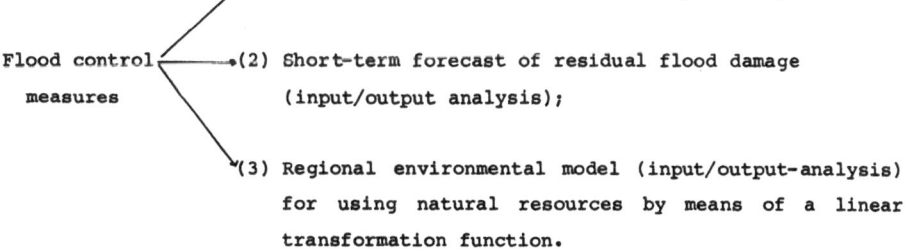

Flood control measures

(1) Regional economic model (input-output analysis);

(2) Short-term forecast of residual flood damage (input/output analysis);

(3) Regional environmental model (input/output-analysis) for using natural resources by means of a linear transformation function.

Figure 2.4. Analysis steps of the economic, environmental and ecological impacts.

(1) The primary economic impact of the project alternative uses an input-output approach. The estimated damages (in monetary terms) caused by a (hypothetical) flood is determined by comparison of the model results with protection against flood and the model results without the project alternative.

The general solution of the regional economic system is:

$$X = (I - A)^{-1} Y, \tag{2.3}$$

with X a vector of economic output, Y a vector of final economic demand and A a matrix of regional coefficients, which is considered to be constant in time. The hydrologic - (regional) economic linkages are assumed to be linear in nature.

(2) The residual flood damage short-term forecasts are obtained in the following way:

$$P'' = P'(I-A')^{-1} Y'', \tag{2.4}$$

with: P' = the present matrix of residual flood damages;

P'' = the (short-term) forecast matrix of residual flood damages;

A' = the present matrix of (fixed) regional coefficients;

Y'' = the (short-term) forecast vector of final demand.

The matrix P'' is of order 6xn, with n the number of regions classified by 6 environmental parameters (viz. vegetation, soil, flood magnitude, sediment, water quality and land-use changes). The matrix A' has coefficients which are considered to be constant in the short term.

(3) The regional environmental module estimates the ecological capacity of the environmental module on an interregional-intersectoral basis, with 6 regions which are part of the Arkansas area. The sectors consist of the 6 environmental elements. The estimation procedure again makes use of the input-output approach. The environmental requirement per dollar of output for each environmental element could have been computed if data about environmental use coefficients whould have been available. The environmental requirements make use of the vector of total environmental output multiplied by environmental use coefficients. Unfortunately, no meaningful biological data were available.

A main advantage of the input-output analysis applied in this model is the availability of a coherent analytical tool to analyse the regional impacts of a project alternative.

Some disadvantages of the input-output approach are:

(i)   the assumption of constant coefficients over time, which means that only short-term forecasts can be made;

(ii) the assumption of homogeneous output;

(iii) the assumption that firms react to increased demand by a rise of output but do not react to a rise of prices.

Because of the absence of meaningful biological data, it has not been possible to estimate the effect of flood damage on bird populations. However, the input-output approach has been illustrated in the environmental module by making use of a hypothetical case of two alternative (with-without) flood protection policies.

2.7. AN ECONOMIC-ECOLOGICAL MODEL FOR LAND-MARINE INTEGRATED DEVELOPMENT

In order to analyse land-marine developments simultaneously, an integrated model with both economic and ecological aspects is developed for the Seto Inland-Sea area in Japan. The aim of the model operationalization is the analysis of the interactions between land and marine development (Ikeda, 1984; Nishikawa et al., 1980). The area has been subdivided by the modellers for that reason into coastal regions and inland-sea regions.
The main characteristics of the area are its rapid industrialization and urbanization during the final quarter of this century. It is a very important area for fishery activities. Serious problems occurred in recent years (e.g., mass death of fishes) because of eutrophication in water. The area also receives inflows of nutrients such as nitrogen and phosphorus from industrial development and urbanization in regions close to the Seto-Inland area. Such nutrients lead to marine eutrophication.
A number of models has been developed in Japan in which both economic and ecological impacts are included: for example the relationship between nutrient inflows and its impact on eutrophication. The main aim of such models was the development of a methodological tool to represent the impacts in monetary terms. The present integrated model will also include factors which cannot be measured in economic measures like monetary terms. An example of a non-economic factor included in the analysis is eutrophication in aquatic environmental living organisms, which is part of the food chain. The model includes the dynamic nature of the eutrophication process as well as socio- economic change and makes use of the nonlinear nature of the behaviour of the modules.
Model tools which have been used here are linear programming methods, input-output analysis and systems dynamics. The overall model structure consists of a socio-economic system and a marine ecosystem; the two systems are based on four interrelated modules, viz. a socio-economic module, a coastal resources demand module, a pollutant emission module and a marine ecological module. The model structure is visualized in Figure 2.5.

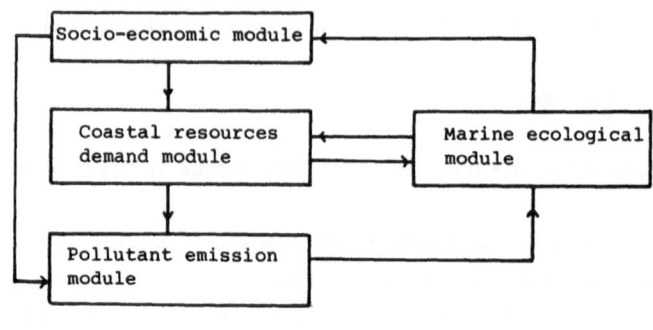

Socio-economic system   Marine ecosystem

Figure 2.5. Structure of an integrated economic-ecological model in Japan.

The basic characteristics of the model are:

- nonlinearity of the behaviour of the system, because of the dynamic nature of the ecological processes. The socio-economic system is for- mulated in linear terms;

- interactions between the socio-economic system and the marine ecosys- tem. The interactions are represented by only a relatively small num- ber of elements.

Interactions between both systems exist in the pollutant emission module and the coastal resources demand module, in the following way:

(a) inflow of nutrients and pollutants such as spilled oil and COD (chemical oxygen demand);

(b) physical destruction of the coastal zone because of land reclama- tion and dredging for the extension of land area;

(c) harvest of marine products from fishing and farming;

(d) recreational use of coastal resources.

The socio-economic activities are restricted by the ecosystem impacts. No direct relationship exists between both parts because human control and in- tervention is allowed. The main characteristics of the four modules from Figure 2.5 as well as the links between them will now briefly be discussed.

(1) The marine ecological module is a systems decision model. The dynamic behaviour is described (in general terms) by differential equations in the following way:

$$\dot{E} = \frac{dE}{dt} = f(E, X, t) \tag{2.5}$$

with $(E, X) = (E_0, X_0)$ (fixed exogenously at $t = 0$), where E and X

are, successively, the state and input parameter vector for all species. If the unit time interval is one year, the annual net production, which has a feedback to the socio-economic module, is obtained by subtracting the state parameter vector for period t from that of period t+1, so that:

$$\text{Net production} = E(t+1) - E(t). \tag{2.6}$$

The model input consists of the amount of nutrients (nitrogen and phosphate) and pollutants (e.g., spilled oil) as well as the utilization of the coastal zone.

(2) The socio-economic module is a multiobjective linear programming model (see also Nishikawa et al., 1977) and describes the socio-economic activities. These activities emit pollutants into the sea.

The constraints in the socio-economic module can be subdivided into hard constraints and soft constraints. The hard constraints may consist of the fixed items of the analysis in case of physical capacity or they may be fixed exogenously with land area, water supply and foreign trade. The soft constraints are influenced (strengthened or relaxed) by human intervention such as education, medical care, social welfare, housing or the living environment.

(3) The pollutant emission module is a module which links the socio-economic module and the marine ecologic module by means of the pollution load in the Seto-Inland sea from all sort of economic activities. The sources of pollution are based on population, agricultural production, cultivated fishery, industrial production and precipitation. The pollution loads are the amount of COD, discharged into the sea, the total nitrogen and the total phosphorus loads as well as the quantity of oil spilled. The transformation of pollutant sources into pollution loads makes use of linear functions.

(4) The coastal resources demand module gives estimates of the demand for coast and shoal water area, obtained from the output of the socio-economic module. The demand for coastal resources is subdivided into use for urban development, recreation, cultivated fishery, industry, harbour, fishing port and preservation of nature and scenery. The demand equations are linear in nature.

The model has not yet been applied for policy formulation in the decision-making bodies, but numerical simulations have been made (mainly for parts of the model, because of the large number of parameters in an economic-ecological model at a regional level which have to be identified by data).

A gap still exists, for example, between ecological information of fish stocks at a regional level (its dynamic nature and sustainability), and eco-

nomic information of fishery industry, such as demand for and supply of
fishes by consumption sectors.

2.8. AN ENVIRONMENTAL ECOSYSTEM QUALITY MANAGEMENT MODEL

In this environmental ecosystem quality control model, a nonlinear ecosystem
module for the Lower Delaware River Valley Region is constructed and linked
to an economic module (Kelly and Spofford, 1977, Russell and Spofford, 1977;
Spofford et al., 1976). The main aim of the analysis is to provide agencies
which are in charge of decisions regarding the economy and the natural environment, with information about costs and impacts on the environment due to
various actions.

The analysis consists of three parts:

     (a) a linear programming model of regional residuals;

     (b) an environmental module;

     (c) an environmental evaluation module.

(a) A regional <u>waste management module</u> denotes the impacts, in monetary dimensions as well as in physical dimensions, of alternative waste management strategies. It uses linear and nonlinear programming analysis. The
objective function to be minimized is based on the costs of water management, while the set of constraints represents the structural relationships among the decision variables.

A <u>linear management module</u> can be described in the following way:

$$\text{Minimize} \quad C = c\,W \tag{2.7}$$
$$\text{subject to } A_1 W + A_2 Z \geqslant B$$
$$H\,Z = X$$
$$X \geqslant S$$
$$X,\ W,\ Z \geqslant 0.$$

X represents the (steady-state) spatial distribution of concentrations of
various materials in the environment, and W is the vector of the activity
levels of the processes used to reduce waste water discharge. The levels
of the standards S depend on the environmental quality which have been
selected for the analysis and are assumed to be given.

The aim of the analysis is to determine the waste water discharge level
Z, that minimizes the total regional costs C of the selected waste water
management level and at the same time meet the minimum requirement level
S. The linear environmental module is included in the set of constraints:
the first restriction represents a set of (linear) relationships that

ensure the production levels and the second restriction represents the availability of natural resources.

A nonlinear form of the management module, with a nonlinear specification of the objective function, can be expressed as:

$$\text{Minimize} \quad F = c\,W + P(Z) \qquad\qquad (2.8)$$
$$\text{subject to } A_1 W + A_2 Z \geqslant B$$
$$W, Z \geqslant 0,$$
$$\text{where} \quad P(Z) = \sum_{i=1}^{n} P_i \left[s_i, x_i = h_i(Z) \right].$$

Nonlinear programming had to be used because the ecosystem module, to be discussed in (b) below, consists of nonlinear relationships. The nonlinear environmental module, which was linearized in equation (2.7) by means of $HZ = X$, is included in the objective function in equation (2.8) by means of a penalty function. $P(Z)$ represents the sum of all penalties P_i, and is associated with the standard S. The individual standards s_i are allowed to be violated in equation (2.8), but only at the cost of a high (monetary) penalty to the objective function. The artificial variables in the penalty function do not have a direct physical interpretation in terms of the model expressed in equation (2.8).

Exogenous constraints which are included in the regional waste management module are:

(i)   minimum production requirements of goods (in monetary terms);

(ii) boundary levels of environmental quality denoted by a minimum concentration level of dissolved oxygen, a maximum concentration of algae, and a minimum standard level of fish;

(iii) spatial distribution pattern of the costs implied by both (i) and (ii).

(b) An aquatic ecosystem module is a simulation model with differential equations describing the rates of change in state variables. The results from (a) become input in the ecosystem module. Because of the regional waste management module is static, only steady state solutions of the aquatic ecosystem module are of interest.

The ecosystem module is an exploratory module with data obtained by aggregation from different locations. The means of data gathering for the aquatic ecosystem module reflect the two constraints of this module. The first restriction was that data have been used, which were collected

before the model operationalization on the Delaware Estuary started. The
second restriction was that the ecosystem module should be used within an
optimization framework like in equation (2.8).

(c) Costs and environmental quality are the main components of the regional
waste management module, and the output of the management module consists
of a vector of residuals classified by substance and location.

The residuals from the nonlinear programming analysis in (a) become input
in the aquatic ecosystem module of the Lower Delaware Estuary and the
regional dispersion concentration analysis. Quantity levels in the regional environment (e.g., SO_2-concentration) are used as input for the
environmental "evaluation" module. Such concentrations are related to
exogenously defined environmental standards.

An iterative procedure between the three modules is then used to select a
vector of residuals that meet the constraints at minimum regional costs. This
approach uses nonlinear programming analysis.

The inclusion of the nonlinear aquatic ecosystem module gives a nonlinear
programming algorithm. However, the usefulness of the nonlinear nature of the
ecosystem module is doubted by the modellers themselves for the Lower Delaware Valley situation, because of the following reasons:

(i)   complexity of economic activities;

(ii)   the fact that algae are not a major water quality problem in the
        study area;

(iii)   computational difficulties with nonlinear programming that increase
        more than proportionally with model size.

The modellers suggest for these reasons and given the state-of-the-art in
ecosystem model development, use of linear models in the optimization framework.

The ecological module, which was originally used for simulation, can also be
incorporated as mentioned above, within an optimization framework. However,
the authors also mention that "lack of appropriate data remains the major
limiting factor in constructing and validating ecological models to be used
for management purposes" (Kelly and Spofford, 1977, p. 442).

2.9. A TRIPLE LAYER APPROACH TO A NATIONAL-REGIONAL ECONOMIC-ENVIRONMENTAL-EMPLOYMENT MODEL

The aim of the study is to analyse economic, environmental and other developments (e.g., energy, employment) in a spatial system at a macro- and meso-
level. An operational national-regional environmental model involving these

issues for the Netherlands has been developed by Hafkamp (1984). This model
is based on a so-called multi-layer projection: it provides a methodology for
combining different systems modules by means of such a multi-layer projection.

A multi-layer projection means that a specific part of a system, as an abstraction of reality, is projected on two or more parallel layers (e.g., economy, environment, energy). Such layers represent structured dimensions of
reality which are taken in conjunction with spatial and temporal aspects.

An integrated national-regional economic-employment-environmental model -
based on three parallel layers - is operationalized for the Netherlands.
Multiregional aspects of the layers are also included, because the country
has been subdivided into 5 regions. The Triple Layer Model (or, shortly, TLM)
is an operational model that integrates economic, environmental and socio-political aspects. The socio-political aspects in this model concern mainly
employment.

As a final result, the multidimensional reality is projected on three mutually interacting parallel layers (economy, environment and employment), with
national and regional links. Relationships are then defined for variables in
the same layer (intra-layer relationships), while also relationships are
defined for variables between different layers (inter-layer relationships).
An example of an inter-layer relationship is the influence of environmental
pollution (part of the environmental module) on the consumption process (part
of the economic module). The modules of the triple layer concept and their
major relationships are given in Figure 2.6 (source Hafkamp, 1984, p. 138).

The model employs a multi-objective decision analysis, based on a set of
different (mutually conflicting) objectives which are influenced by a number
of relevant decision variables. A unique optimum solution cannot be found for
such a conflict situation because of the complexity of the analysis and the
mutual dependence of relevant variables. Priorities have then to be assigned
to the alternative solutions, by using a method of displaced ideals in order
to obtain a multiobjective optimization solution.

The economic module is a national-regional module (incorporating variables
such as production, consumption, imports and investments) which makes use of
a five-region input-output model. The impacts of various alternative economic
policies on the economy are analysed by means of a simulation model.

The employment module gives a description of supply and demand of labour in
all regions and economic sectors.

The environmental module describes emission and diffusion of air pollutants
(SO_2, NO_x and dust particles) from fossil fuels. Water pollution and solid
wastes are not included in the present analysis. The modelling of real world

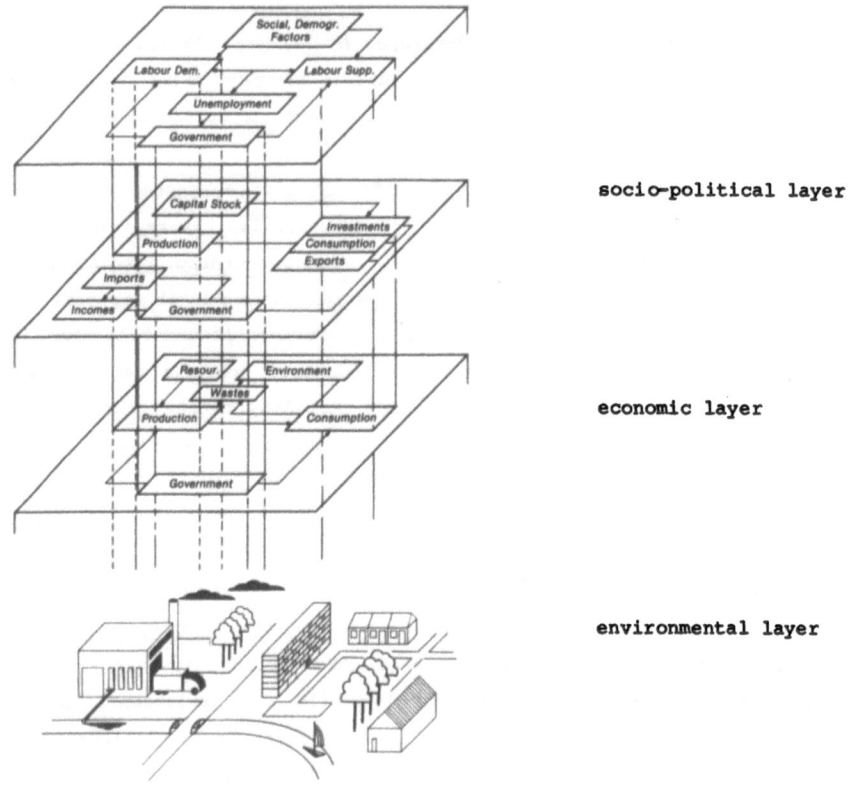

Figure 2.6. The TLM, its modules and their major relationships.

phenomena in the successive separate layers is taking place in their natural units of measurement (including spatial dimensions).

Nonlinear parts of the multi-objective model are linearized for computational reasons. The input-output model is represented by linear functions and is estimated with single-point estimates. The other linearized equations are estimated with ordinary least squares.

The year 1973 is a base year of the analysis and no further time dimensions are included in the multi-objective framework. The regional input-output model was updated from 1970 to 1973, by assuming a temporarily constant regional sectoral activity pattern.

Of course, a multi-layer projection may encompass two, three or more layers, which depends on the specific modelling context and it is essentially based on a modular design of variables and structures of the model concerned.

2.10. SYSTEMS APPROACH OF ECONOMIC-ENVIRONMENTAL-ENERGY ANALYSIS

The island Gotland in Sweden is an agricultural area and was the study object concerning the interactions between energy/economic activities and ecological activities. A systems framework was developed and operationalized in order to provide quantified information concerning the interactions. Emphasis was given to the energy costs of agriculture and the energy flows between the interacting sectors such as industry, agriculture and forestry. Some attempt was also made to assess the environmental costs of a decline in water quality. The economic activities (classified into industry, agriculture and forestry) consume natural resources and fossil and non-fossil fuels, they produce economic output and have impacts on the environment (Zucchetto and Jansson, 1979; 1981).

The modules and their relationships for the Gotland island are denoted in terms of economic feedbacks as well as energy feedbacks in Figure 2.7.

The relationships between both modules in Figure 2.7 are described in energy measures (in terms of joules) at the left hand side and described in economic measures (in money terms) at the right hand side.

The right hand side part of Figure 2.7 consists of an input-output approach and an optimization procedure. The input-output elements of the energy flows are subdivided into both fossil and non-fossil fuels and denote the regional energy flows between a natural resources module and an economic activities module. Forest and fish species are part of the natural resources module. Part of the forest activities is called forestry because it is used in the economic activities module as productive forest area.

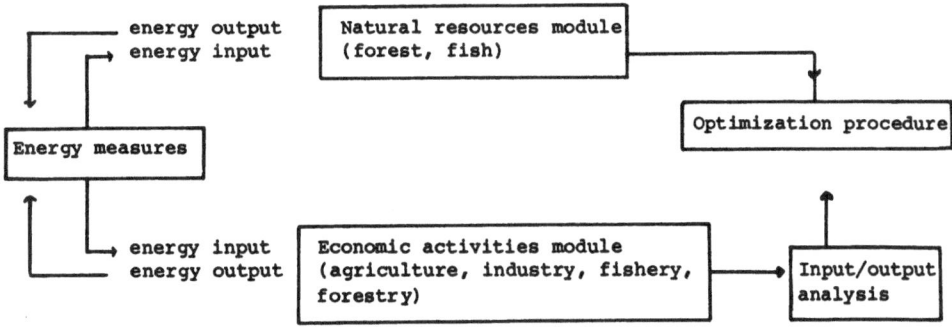

Figure 2.7. Feedback relationships for the Gotland region.

An example of the energy input and output flows regarding harvest in agriculture is the following: low energy level input flows (from among others fuels
by tractors) are transformed during the production process into harvest which
represent high energy equivalents. The entire regional energy system also
includes the linkages between the agricultural system and other components on
the island, such as the use of electricity and fuel oil in farms, and the
input of capital, fertilizer, goods and services, sunlight and number of
man-hours to produce crops and animals.

The economic activities in the region, which consist mainly of agriculture,
industry, fishery and forestry, are linked to each other in monetary terms by
an input-output analysis. The input-output model is part of an optimization
procedure which maximizes, subject to economic and environmental constraints,
the sum of values added for the economic activities, forest and natural
lands, and is of the following type:

Maximize $\quad (V_X \cdot X + V_Z \cdot Z)$ or: Maximize $(V_X \cdot (I-A)^{-1} Y + V_Z \cdot Z)$ $\quad$ (2.9)

subject to: $\qquad X < C_X \qquad\qquad$ subject to: $(I-A)^{-1} \cdot Y < C_X$

$\qquad\qquad\qquad Z < C_Z \qquad\qquad\qquad\qquad\qquad\qquad Z < C_Z$

$\qquad B \cdot X + D \cdot Z < R \qquad\qquad B \cdot (I-A)^{-1} \cdot Y + D \cdot Z < R$

$\qquad\qquad X, Z > 0 \qquad\qquad\qquad\qquad\qquad\qquad\qquad Y, Z > 0$

where: $\quad X$ = vector of output from the economic based sectors;

$\qquad\qquad Y$ = vector of final demand from the economic based sectors;

$\qquad\qquad Z$ = vector of forest and natural lands;

V_X, V_Z = vectors of value added coefficients, respectively for the
economic sectors, and for forest and natural lands;

$\qquad\qquad A$ = matrix of technical coefficients;

C_X, C_Z = matrices of capacity constraints on production and on land
use for forest and natural lands respectively;

$\qquad B, D$ = resource requirements matrices for the economic based sectors,
and for forest and natural lands respectively;

$\qquad\qquad R$ = vector of resource availabilities.

The right-hand side equations of (2.9) are analogous to the left-hand equations, because $X = AX + Y$ is transformed into input-output terms as $X = (I-A)^{-1} Y$.

Different scenarios of energy price developments have been considered in the
model specified in equation (2.9). An example is a rise in the price of one
specific energy type. The substitution of energy types occurs because of
energy price scenarios, and follows from the above mentioned optimization
procedure.

The time horizon of the energy analysis covers the medium long term (about 10

years) and is selected to achieve different energy scenarios and their consequences within the planning horizon.

The feedback between economic activities and the use of natural resources are
denoted in economic terms within the optimization procedure, with the output
of both modules being maximized subject to economic restrictions and natural
resources restrictions. The modules are related to each other in equivalent
dimensions (viz. monetary terms and energy flows).

The spatial dimension covers the whole Gotland region without a further distinction being made for the region.

It is worth noting that this analysis is based on energy flow analysis. In
this context, a quite different approach to the energy flows is described by
Spiller et al. (1981) for agricultural development. A simulation model was
developed concerning the nitrogen flows in the agricultural system and the
environmental effects of such flows on water quality. Differential equations
are given for the quantity of soil water and ground water, as well as the nitrate volumes in soil and in groundwater. The impacts between variables are
displayed by means of a flow chart of the variables. Such a flow chart was
developed from Odum's natural energy systems procedure (see also Odum,
1971).

2.11. SIMULATION OF THE IMPACTS OF HERBICIDES ON THE ENVIRONMENT AND AGRICUL-
TURE

The functioning of ecosystems and socio-economic systems depends on a balance
between constructive and destructive forces. The consequences of destructive
forces for the natural environment and agriculture can be observed convin-
cingly in South-Vietnam, a country whose physical and ecological structure
suffered from a war for many years.

An analog computer analysis is developed to simulate part of the ecosystem
development in this country during the war-period between 1950 and 1970 (see
also Brown, 1977). The simulation includes the effects from bombing and herbicide spraying on ecosystems and cities, as well as changes in different
types of land quality. The model also simulates the effects on the regional
economy and population distribution patterns. Land categories are subdivided
into four components, viz. development of city, agriculture, forest and mangrove. The city component consists of the elements which describe the use of
land area, city structure, goods and money. The natural systems have been
subdivided into agriculture, forest and mangrove; each of them has two components, viz. the use of land area and its biotic structure.

Input of the simulation analysis consists of ordering and disordering factors

in energy elements, and is denoted in terms of materials and energy.
The land and structure characteristics of mangrove, forest, farms and cities
are represented by (non)linear differential equations, like:

$$\dot{L}_1 = \frac{dL_1}{dt} = k_1 L_5 A + k_2 L_5 - k_3 (L_1 + L_2 + L_3 + L_4) W \qquad (2.10)$$

and

$$\dot{S}_1 = \frac{dS_1}{dt} = k_4 L_1 I - k_5 (S_1 + S_2 + S_3) (G + A) L_4 P_2 - k_6 S_1 \qquad (2.11)$$

where: L_1 = mangrove land (in acres);

L_2 = forest land;

L_3 = agricultural land;

L_4 = city land;

L_5 = bare land;

S_1 = mangrove structure (in weight terms);

W = war effort;

I = natural energies;

A = U.S.A.- aid (money terms);

G = imported goods and fuel;

$k_1, \ldots, k_6$ are pertaining parameters.

Equations analogous to (2.10) and (2.11) are specified for forest, agriculture and city components. The land related equations are linear differential
equations and the structure characteristics are described by nonlinear differential equations. Nonlinear differential equations also exist for the
rural and city population variables:

$$\dot{P}_1 = \frac{dP_1}{dt} = k_{21} (k_{14} L_5 + k_{16} L_5 A) k_{32} P_2 - k_{23} P_1 W - k_{24} P_1 \qquad (2.12)$$

where: P_1 = rural population;

P_2 = city population.

Nonlinearity in the variables exists in equation (2.12) because the rate of
change of variable P_1 depends among others on variable $L_5 \times P_2$.

The variables are represented in their natural units of measurement, which means that the land variable is measured in terms of acres, the variable which denotes the structure by weight terms, the population variable by number of people, and aid and war efforts in monetary terms.

Two remarks are in order here, because of their effects on the reliability of model outcomes:

(i) the assumption has been made that future distribution patterns in the country are similar to the existing patterns: for example the constant distribution in time of aid between urban and rural areas.

(ii) "the model is dependent on data and relationships that are not known with precision" (Brown, 1977, p. 411).

Direct links between economic phenomena (in terms of monetary aid) and ecological processes (natural systems as agriculture, forest and mangrove) are modeled in a system of nonlinear differential equations. The model output indicates the rate of change of the natural systems in terms of structure and land. Different scenarios of U.S. aid are simulated for the period 1970-2000, viz. constant, increasing and decreasing aid by the USA.

2.12. ECONOMIC-ECOLOGICAL ANALYSIS BY SIMULATION AND OPTIMIZATION

A study is operationalized in Canada to analyse the relationship between man and nature (Lonergan, 1981). The integration of economic and ecological systems initially resulted in an input-output framework to characterize the interrelationships. However, disadvantages in using input-output in such multidisciplinary efforts include "measurement problems, unit disparities and the fixed input coefficient assumption" as well as "the absolute monetary evaluation" (Lonergan, 1981, p. 118). The author proposed for this reason another analytical tool to link an ecosystem simulation module with an economic optimization module for regional land-use planning. The formal mathematical representation becomes the following:

$$\text{Maximize} \quad c^T \cdot x \tag{2.13}$$

$$\text{Subject to: } Ax \leq b$$
$$x \geq 0$$
$$\dot{x} = \frac{dx}{dt} = Dx + Ez$$

where:
x = vector of main variables (animals);

z = vector of driving forces in the simulation analysis;

c, A, D, E = vector and matrices of constant coefficients;

b = vector of constraints.

The simulation-optimization approach is applied to the Choptank River watershed in Maryland (Canada), including the Chesapeake Bay, ecosystems to study economic-ecological conflicts. The analysis makes use of the approaches developed by Kelly and Spofford (1977), in an analogous way as equations (2.7) and (2.8) in Section 2.8.

Two of the primary factors in the area which contribute to the decline of aquatic vegetation are the increasing population in the watershed and the use of herbicides in agriculture. A conflicting mutual dependency exists between the land based system with agriculture as a major component and the aquatic system with commercial and sport fisheries. Agriculture and fishery both provide a contribution to the regional economy.

The simulation module is based on the impacts between the state variables of the aquatic system for the Choptank River. The impacts between the state variables are given in a systems diagram (see Lonergan, 1981, p. 124) and make use of the energy systems language of H. Odum. First-order differential equations are used to describe the aquatic system.

This module has been calibrated in both physical and energy units as well as monetary units in order to estimate the effects of different value metrics on the systems interpretation. A monetary criterion of value was used by Lonergan, with physical outputs of the ecological sectors to be converted into monetary terms, by using market prices (e.g., commercial fisheries) and shadow prices (e.g., sport fisheries). The result of this monetarization is used in the economic optimization procedure.

The model can also be reformulated in terms of energy flows with an objective function that maximizes energy efficiency.

Such approaches with different units of measurement (either in monetary or in physical dimensions) show that no one-to-one correspondency exists between the aim of a research question raised and the analytical tools used. Any research question can be analysed in various ways, while using different unit measures which may lead to dissimilar conclusions.

The optimization module consists of a monetary value structure with an objective function that maximizes the value added from various sectors in the regional economy. The objective function consists of three parts:

(1) value added (in monetary terms) of the four activity sectors, viz. agriculture, commercial fisheries, sport fisheries and industry, with time horizon t and a discount rate to monetarize future income;

(2) costs of the economic activities that are external in nature to the specific sectors (e.g. pollution abatement strategies) with time horizon t and a discount rate to monetarize future costs;

(3) social costs of a declining activity sector when, for example, a

particular activity is decreasing in time.

The model constraints of the objective function are:

(i) the number of acres allocated to an agricultural component in some year, which is equal to the number of acres in a previous year multiplied by a weight component. The weight chosen consists of the mean crop price in a previous year divided by an average price of crop over the past 20 years;

(ii) upper bound on the amount of land available for agricultural use;

(iii) upper bound on the harvest of fish;

(iv) production function of the total output of some activity component in the activity sector. The components from the economic sectors are agricultural output (as a function of herbicide and fertilizer), and fish population (as an indirect function of herbicide, fertilizer and sediment runoff, obtained from the simulation module);

(v) non-negative output levels.

The simulation module and the optimization module are linked to each other in an iterative two-step approach as mentioned above.

The methodological tool developed in the study mainly deals with the linkage of simulation and optimization modules. Both approaches are rather flexible: a simulation module can be updated as additional empirical results become available and an optimization module can be altered by substitution of objective functions and the addition (or deletion) of constraints.

2.13. URBANIZATION AND ENVIRONMENTAL PLANNING AND DESIGN

In this model on urbanization and environmental planning, consequences and the effectiveness of different planning strategies for the south-eastern part of the Boston area in the U.S.A. are analysed. The development of the area is characterized by a rapid suburbanization in the major number of cities. The process of suburbanization started during the mid-sixties.

In order to analyse in a comprehensive way all relevant urban environmental impacts a large number of computer modules (28 in total) have been developed and operationalized by a multidisciplinary research team with regional planners, economists, ecologists and environmental engineers to obtain the consequences of suburban growth (see also Steinitz et al., 1976, Steinitz and Brown,1981). The computerized data analysis is dealt with at three levels:

(i) data base generation, which consists of only the basic data manipulations such as retrieval and updating;

(ii) allocation modelling and evaluation modelling. The 28 modules are modelled separately and are linked to each other in an artificial way by means of a data base system;

(iii)   interactive control routines to link the individual modules.

The aims of the model operationalization may be different in nature, viz.:

(a) pre-analysis which gives a description of regional characteristics when no model will be used.

(b) single module analysis when the modules are developed individually to answer specific questions.

(c) project evaluation when the impacts of a selected project with corresponding land-use allocation are determined by means of the appropriate evaluation modules.

(d) plan evaluation which requires land use allocation in an analogous way as in the project evaluation, with the exception that now all land use types are specified.

(e) sensitivity analysis which requires different simulation runs of one or more modules. The values of one parameter or the assumptions concerning the impacts are changed during each simulation run. The difference between simulation runs indicate the sensitivity of the modules due to the changes in parameter values.

(f) legal testing when the consequences of changed laws, statutes or administrative procedures are determined.

(g) gaming when various planning strategies are developed. This makes use of an interactive computer modelling process.

(h) alternative strategy simulation when changes in land use are allocated according to a mixture of planning strategies. The impacts of a selected alternative are evaluated here for all modules.

An example of an alternative strategy simulation is a trend projection of future regional development, based on current policies. A temporal resolution of five years is selected for a cycle of allocation and evaluation because a five years period corresponds to the view of local and regional planning. The temporal coverage of the alternative strategy simulation is the period 1975-1985.

The means of model operationalization mentioned above show that a module can be operated individually when only one aspect of urbanization is relevant. The modules can be linked to each other when mutually dependent and conflicting aspects of urbanization are to be analysed.

Various levels of spatial aggregation may also be dealt with, viz., a region, a town as well as other functional zones. The choice of the spatial aggregation level depends on the users needs.

The modules are subdivided into allocation modules and evaluation modules.

The allocation modules provide a particular type of land use to a location in the area and such model outcomes are used for updating of the base year conditions from the data files. The allocation modules deal with the allocation of housing, industry, public institutions, recreation, conservation and solid waste.

The update of data files is obtained from the allocation modules and will be used as input for the evaluation modules. The evaluation modules describe the impact on the environment, demography and tax institutions of the new land-use allocations and they consist of modules which are related to soil, water use and water quality, vegetation of natural change, air quality and noise quality. Household demand of water, for example, is estimated as a function of median family income and metering practices.

The emissions of sulfur dioxide and carbon monoxide are translated into (regional) concentrations by a Gaussian plume dispersion equation. Such concentrations are evaluated by means of a comparison with emission standards set by the Environmental Protection Agency (EPA) in the USA.

Each module has a spatial and temporal resolution which is appropriate for its specific function. The housing module, for example, has a spatial resolution of one hectare, while the air quality module has a more aggregated spatial resolution with areas of three square kilometers. The vegetation module has a temporal resolution of one day.

The study showed the relevance of developing systems models as an instrument in various aspects of land use planning by means of an interactive development process and critical discussions between analysts and the users of a model.

2.14. AN INTEGRATED REGIONAL ENVIRONMENTAL MODEL FOR PHYSICAL PLANNING

An integrated environmental model (IEM) has been operationalized for spatial planning purposes at a regional level, based on a medium to long term time horizon (a temporal coverage of 10 to 15 years forecasts). A case study to develop an IEM has been carried out for the West-Brabant region in the Netherlands (Arntzen et al., 1981, Brouwer et al., 1983). The environmental and spatial aspects of an extension scheme in regional physical planning are presented in relation to the direct and indirect economic, demographic and facility impacts.

The analysis consists of five modules which are linked to each other, viz. a demographic, an economic, an ecological, an artificial environment and an intermediate module. The links between the modules are represented in Figure 2.8.

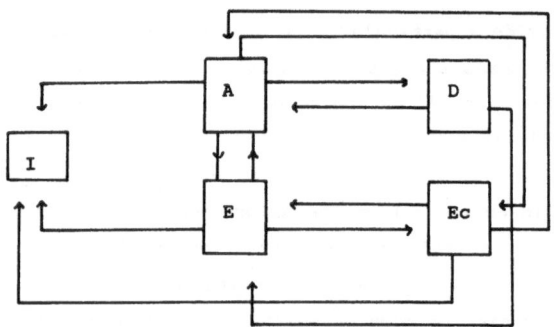

D = demographic module   A = artificial environment module
E = economic module   I = intermediate module
Ec = ecological module

Figure 2.8. Representation of the interactions between modules.

The main characteristics of the modules will be described below briefly:

(1) The demographic module denotes the composition of population, classified by age and sex, and makes use of the cohort survival method (a
scenario approach of expected future population development).

(2) The economic module also makes use of a scenario approach to determine the demand for labour classified by the economic sectors. The
labour demand equations are linear in nature. The supply of labour in
the region is obtained from the population variable in the demographic module.

(3) The artificial environment module or facilities module makes use of
linear equations to determine the demand for houses, the use of water
(by households as well as firms), the demand for recreational facilities, supply of garbage (from households and firms) and also the
pollution by phosphate, nitrate, sulfur dioxide and nitrogen oxides.

(4) The ecological module is a simulation model with non-linear difference equations and consists among others of quantity of algae in water,
quantity of aquatic macrophytes and concentration of salt. The parameters to quantify the relationships are based on literature surveys
and model fitting.

(5) The intermediate module determines a balance between supply and demand of land use classification. The emissions of sulfur dioxide,
obtained from the artificial environment module, is an input to a
dispersion model in the intermediate module to locate the concentration of sulfur dioxide.

The interactions between modules in Figure 2.8 are described at a more aggregated level than the relationships within a module at the level of individual variables. The size of population - a variable in the demographic module - is related directly to the supply of labour in the economic module and is also related to the household variables which are part of the artificial environment module. A mutual dependence exists between the ecological module and the artificial environment module: the level of recreational activities is an independent variable in the ecological equations which specifies the quantity of phosphate and fish. When these ecological variables exceed a limit level (in terms of g/m^3) in some year, the recreation level will decrease in the next year by 25%.

The economic and ecological modules are linked to each other partly in a quantitative way, partly in a qualitative way due to lack of data on the systems interactions. The emissions of sulfur dioxide are determined by the level of employment multiplied by emission factors.

All modules, except the ecological module, have a temporal resolution of one year. The level of the ecological variables is determined for 3 months periods.

Data have been collected from literature survey for the ecological processes, and from regional economic-technological institutes and the national bureau of statistics for the other activities.

Figure 2.8 shows that the ecological module has direct links with the three other modules. Consider for example the direct relationship between the artificial environment module and the ecological module where a mutual dependence is operationalized between recreational activities and the phosphate concentration in water.

2.15. INTERACTIONS BETWEEN ECONOMIC-SOCIAL-ECOLOGICAL SYSTEMS IN A REGION OF INTENSIVE AGRICULTURE

An UNESCO research program called Man and Biosphere (shortly, MAB) was initiated in 1971 to build a research bridge between the ecological, economic and social sciences. A multidisciplinary research project started also in the F.R.G. within the framework of the UNESCO program. The research background covered the fields of ecology, social science, law and geography (see also MAB-vol. 7, 1982; MAB-vol. 14, 1983; Müller, 1985).

The aim of the German project was to elaborate a methodology for linking ecological, economic and social systems with each other into an interactive model by means of a case study example in the South Oldenburg region.

The South Oldenburg region was selected because it is an intensively cultivated agricultural area. Such agricultural activities have impacts on ecological, economic and social phenomena which can be analysed in an integrated approach. Additional advantages to select this region were the possibility of a cooperation between a number of neighbouring universities and the availability of results from preparatory work which had already been undertaken (e.g., extensive data collection).

The agro-system consists of an ecological module, an economic module and a socio-political module. The model structure is represented in Figure 2.9.

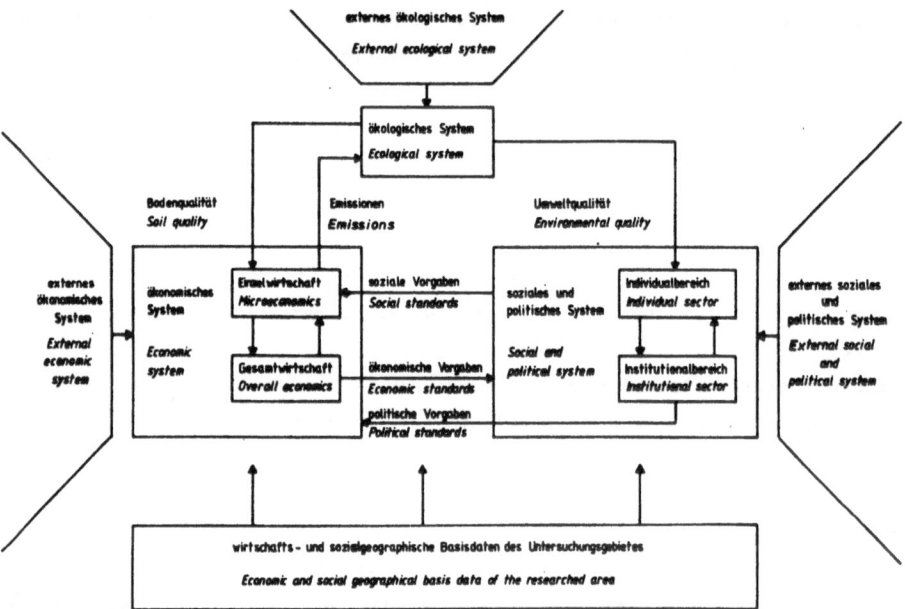

Figure 2.9. Structure of the overall agro-ecosystem
(Source: MAB-vol. 7, 1982, p. 81).

The South Oldenburg region is one of the areas in Europe with extremely high animal waste application which creates ecological problems (high NO_x-levels in ground water used for drinking water supply). The environmental problems with the high nitrate levels in groundwater of the Vechta district are depicted in Figure 2.9. This area is part of the South Oldenburg region in the F.R.G.

The most significant research objective deals with the methodological tools to link the modules. The module structure demonstrates the economic relevance of the farm enterprises as well as the ecological consequences of farming. The analysis was initiated therefore with the development of a simulation model for agriculture based on intensive animal keeping (at the micro-economic level). This shows the direct impact of the economic module upon the ecological module. A large quantity of liquid manure is used on the fields as natural fertilizer. Problems arise from overfertilization during dry periods because of high levels of nitrate into the groundwater. For that reason, neither the market, nor capital nor labour but the stress which is put on the environment becomes the limiting factor for agricultural production. The use of liquid manure as fertilizer can be reduced only at the cost of either expanded storage facilities of liquid manure or the possibilities of export to other countries.

The economic module consists of an analysis for an agricultural enterprise with animal batteries at the micro-economic level. The aim is the simulation of micro-economic decision making processes and behaviour of a farm as well as the simulation of the resulting economic and ecological consequences. The model of agricultural structure and of the total regional economy, aggregated from the micro-economic simulation model, is developed to quantify the influence of environmental protection upon the regional economy.

The macro-economic model which covers the South Oldenburg region consists of variables which are relevant for micro-economic as well as political decisions.

The ecological module will describe the liquid manure effects on subsystems (for example, soil, groundwater, surface water, flora and fauna) as well as the feedback to the socio-economic module at the regional level.

Because of the difficulty in operationalizing an ecosystem, the ecological module is restricted here to an agro-ecosystem which is dominated by one driving force, viz. the excessive amount of animal excrements to be disposed of. This single driving force should yield significant insight into the structure and function of an agro-ecosystem in the region. The impact of liquid manure on the nitrate concentration of the groundwater is covered by the micromodel of the agro-ecosystem.

The socio-political module investigates and formulates the behaviour of the people involved who are living in the project region. This module describes the public concern about the nitrate levels in groundwater and its impact on human health, as well as the agricultural interest groups who are against the limitation of liquid manure use.

One detailed model of a single agricultural enterprise is finished so far. The main part of the modelling efforts is still underway and depends primarily on the possibilities to raise funds.

A couple of coordination problems which might be relevant in multidisciplinary projects are discussed briefly below (see also MAB-vol. 7, 1982):

(i) practical coordination problems:

- scientists should be qualified to cooperate with other disciplines

(e.g., a social scientist needs a statistical background in order to cope with approaches used in natural sciences). A coordination of personnel is relevant for that reason.
- results obtained from one project (e.g., development of a module) serve as input for another project, which makes a time oriented coordination necessary.

(ii) <u>a coordination with regard to different contents</u>:
- a coordination of methodological tools from different disciplines is essential because of:
  (a) different levels of measurement of variables. In natural science the majority of variables is measured in metric units; in the social sciences often nominal or ordinal measured variables are available.
  (b) difference in time structures. The natural science mainly deal with a continuous time structure, while the social sciences mainly deal with discrete temporal scales.
  (c) differences in the means of data gathering, either by controlled experiments in the natural science or by surveys which is the main data source in the social sciences.
- theoretical coordination. The interface between different sub-projects should be available in such a way that the output of one sub-project is available as input to another sub-project.

Organisational aid and supporting techniques are necessary to cope with the above mentioned problems and "there is a need that all subprojects use the same systems concept" (MAB-vol. 7, 1982, p. 101), so that appropriate methodological tools are available to cope with the different model approaches.

2.16. CONCLUDING REMARKS

A wide variety of models, instruments and techniques - developed during the last ten years - have been discussed in this chapter to describe environmental phenomena for the analysis of environmental policy issues such as land use planning, water resource planning, and recreational planning. The models discussed differ in their design and focus of application and they may deal with all types of the environmental categories such as water, air and land use. All these environmental phenomena can in principle be operationalized by means of integrated environmental models.

The presentation and discussion of the survey of models is the frame of reference with respect to this study. The design and tools of the IEMs will be further elaborated in the remaining part.

While the models discussed vary in their design and focus of application, they also have characteristics in common. First, all of them have direct or indirect feedbacks between an economic and an ecological module because they include economic phenomena as well as ecological processes. A set of interrelated variables are expressed in terms of modules. Second, the spatial dimension of all modules covers the regional level, which is a spatial entity between the local level and the national level.

A summary of common characteristics of the IEMs, dealing with the aims of the study and the analysis, the modules and mathematical tools used, the spatial and temporal dimension as well as the dimension of the variables and the availability of information, will be evaluated in the following chapter. The evaluation of the above mentioned points of reference give a strength/weakness analysis of the models discussed in the survey.

CHAPTER 3. EVALUATION OF INTEGRATED ENVIRONMENTAL MODELLING APPROACHES

3.1. INTRODUCTION

In the late sixties and early seventies, regional economics mainly dealt with economic-based tools such as, input-output analysis, gravity/entropy models, shift-share analysis, and cohort survival analysis (see also a survey on the state-of-the-art of regional economic research by Richardson, 1978). However, the conflicting nature between environmental aspects and economic phenomena became a field of interest in regional-economic and environmental policy making in the same period, because of the growing concern about the rise in complexity of conflicting public planning and choice problems in our society. A lot of national and regional input-output models has been developed in that period to link economic and environmental systems. Environmental pollution, for example, has been integrated within the input-output approach by means of the determination of pollution coefficients - with either fixed or varying levels for the industrial sectors - with regard to total pollution (see among others the classical tools developed by Isard (1968), which have been extended to interregional flows and pollutants, as well as other environmental factors by among others Lesuis et al. (1980)). The input-output (I/O) approaches, which are developed by Cumberland, Leontief and Isard, and which are based on static equilibrium principles, provide a consistent accounting system. However, they are not adequately suited for the analysis of concrete environmental policy issues (such as land use planning, water resource planning, recreational planning or urban renewal), because the use of regional environmental I/O models has some constraints, such as the static equilibrium assumption, the huge effort necessary to collect all relevant data, and the lack of opportunity to analyse intra-regional capabilities (Solomon and Rubin, 1985).

The interrelationships between economic, demographic, physical environmental factors and natural environmental aspects have been recognized and operationalized in regional impact studies from the mid-seventies and later (Solomon, 1985). The fourteen models discussed in Chapter 2 dealt with various types of integrated environmental analyses. A distinction was made in the survey between the main environmental categories air, water and land use. The representative sample of situations in which modelling approaches have been developed shows the major progress during the last ten years in modelling the environmental impacts of regional development projects, as well as the state-of-the-art in this area of modelling tools.

The main elements of and the correspondence between the models in Chapter 2 will be evaluated below, and be summarized in Table 3.1. Nine points will be emphasized column-wise in that table to characterize the models in the survey from Chapter 2:

(i)   the <u>objective of the study</u>. The objectives of the modelling exercise may deal either with analytical/methodological/academic use, or with policy/management use as an instrument for policy makers;

(ii)   the <u>objective of the analysis</u>. The objectives of the analysis may vary from the development only of a data base, to the exploration or description of phenomena as well as to an explanatory or a forecasting model;

(iii) the <u>modules</u> included in the model, which may focus on (the interactions between) economic, environmental, energy, and demographic aspects among others;

(iv)   the <u>mathematical tools</u> used to operationalize the concept of the modules;

(v)   the <u>spatial dimension</u> of the study, which may vary from a single-regional analysis to an interregional-national analysis. As already mentioned in Section 2.1 a region is simply defined to be a spatial entity which varies between the local level and the national level;

(vi)   the <u>temporal dimension</u> of the study. The temporal resolution may be different for the various modules, varying from a daily basis to an annual change focus. The temporal horizon of the study may also change from an annual basis for short-term impact studies to a medium or long-term (at least five to ten years predictions) period for public planning and regional impact studies. A model is static in nature when no temporal dimension is included in the analysis;

(vii) the <u>dimensions</u> of the variables, which may be depicted in, for example, monetary terms, count terms, or energy flows. The variables may also be described in dimensionless figures;

(viii) the <u>availability of information</u>, where a distinction will be made between the source of information, and the precision of information for the various modules;

(ix)   the <u>model evaluation</u> by means of their strong and weak points. Such a strength/weakness analysis deals with the main objectives and the aims mentioned in the previous points.

The nine points mentioned will be discussed briefly in Sections 3.2 to 3.5. The two objectives categories of the IEMs as mentioned in Table 3.1, viz. the objectives of the study (either methodological development or development for policy use), and the objectives of the analysis (i.e., data base generation,

Table 3.1. Evaluation of IEMs.

(Section) Model	Objective of study	Objective of analysis	Modules	Mathematical tools	Spatial dimension	Temporal dimension
2.2	methodology	evaluation	economy, natural environment	optimization (compromise programming)	one region	static
2.3	policy use	forecasting	economy, natural environment, regional planning	simulation (nonlinear differential equations)	one region	a thirty years horizon
2.4	policy use, methodology	description, explanation, forecasting, evaluation	economy, natural environment, water management	scenario analysis, optimiza- tion, evaluation with a scorecard	a country with 77 regions	a ten years horizon
2.5	policy use	description, evaluation	economy, natural environment, water resources, regional planning	scorecard to represent policy alternatives	one region	static
2.6	methodology	description, explanation	economy, natural environment, regional wa- terplanning	input-output	one region	static
2.7	policy use	explanation, forecasting	economy, water resources, emissions, natural environment	linear programming, input-output simulation	one region	one year time steps
2.8	policy use	explanation, forecasting	economy, natural environment, water resources	linear programming, linear differential equations	57 subregions in Lower Delaware River	static (steady state)

Table 3.1. (continued).

(Section) Model	Objective of study	Objective of analysis	Modules	Mathematical tools	Spatial dimension	Temporal dimension
2.9	methodology	explanation	economy, natural environment	optimization (multi-objective analysis)	a country with five regions	static
2.10	methodology	description, explanation	economy, natural resources	optimization input-output	one region	static
2.11	policy use	forecasting	economy, natural environment, regional planning	scenario analysis, simulation (nonlinear differential equations)	one region	a thirty years horizon with one year steps
2.12	policy use	description, forecasting	economy, natural environment, regional planning	optimization simulation (linear differential equations)	one region	one year time steps
2.13	methodology	data base, description, explanation, forecasting	economy, natural environment, regional planning	linear regression, dispersion model	one region	time steps to be chosen
2.14	methodology	description, forecasting	economy, natural environment, demography, regional planning	scenario analysis, simulation (difference equations)	one region	fifteen years time horizon with one year time step; a quarter time step of the natural environment module
2.15	methodology	description, explanation	economy, natural environment, agricultural planning	systems dynamics	one region	static

Table 3.1. (continued).

(Section) Model	Dimension of variables	Availability of information	Strong points	Weak points
2.2	dimensionless figures	no comments	simple tool to evaluate economic and environmental phenomena in an optimization framework	variables are transformed into dimensionless figures
2.3	monetary and energy dimension	use of historical data (the external forcing function is considered to be constant)	simulation analysis to include the impacts between economic, environmental and planning phenomena	constant level of the external forcing function is considered in time
2.4	natural dimension of measurement	data are of poor quality for the environment module	development of a methodology for water management analysis	direct relationships between economic activities and ecological impacts are ignored
2.5	natural units of measurement	no sampled data are available for the ecological impacts	development of a useful tool to represent policy alternatives	the interaction between economic and ecological phenomena is not included
2.6	monetary dimension	no comments	economic aspects, residual flood damage, and environmental processes in an I/O-model	(multi) regional input-output modelling needs a lot of information
2.7	natural dimension of measurement	no comments	integration of (socio-) economic, ecological resources and pollution emission aspects	a lack of information still exists on the dynamic nature and sustainability of fish stocks
2.8	monetary and physical dimension	lack of appropriate data for ecological impacts	waste management activities included in an aquatic ecosystem module	a lack of information exists to construct ecological models for management purposes

Table 3.1. (continued).

(Section) Model	Dimension of variables	Availability of information	Strong points	Weak points
2.9	monetary and energy dimension	no comments	representation of economic, environmental and energy phenomena by means of interrelated layers	no ecological processes are included in this stage of the TLM model
2.10	monetary and energy dimension	no comments	input-output analysis included in an optimization framework	emphasis upon monetary evaluation of activities
2.11	monetary dimension; acres for land-use types	information not known with precision	forestry, agriculture and urban development in a simulation approach	lack of information concerning the the impacts between variables
2.12	monetary and energy dimension	no comments	integration of economic and ecological phenomena	optimization procedure only includes a monetary value structure
2.13	natural dimension of measurement	no comments	objectives of modelling exercise depend on level of regional planning (e.g. project or alternative strategies)	the 28 modules have been developed independently from each other
2.14	natural dimension of measurement	information from offices of statistics and from literature. Quantitative and qualitative impacts	economic, geographical and spatial factors are analysed in relation to the natural environment	data constraints limit the scope of the modules
2.15	natural dimension of measurement	different ways of data collection and different levels of measurement of the variables	integrated focus of economic, environmental and political aspects for agricultural activities	analysis of only micro-economic activities are finished. The main part is still underway

description, explanation, forecasting, or evaluation), are discussed in Section 3.2. The modules included in the models from the survey, and the mathematical tools used to operationalize the modules are discussed in Section 3.3. A cross-classification concerning two modules (the economic module and the natural environment module), according to the mathematical tools used is also presented in that section.

The dimensions of space and time of the models are discussed in Section 3.4, which shows the difference in spatial and temporal dimensions to operationalize the economic module and the natural environment module.

The evaluation on the availability of information in Table 3.1 showed that a lack of appropriate information leading to difficulties in some IEMs was mentioned. This will be discussed briefly in Section 3.5.

This chapter will be concluded in Section 3.6 with some concluding and final remarks on the review and evaluation of the IEMs from Chapters 2 and 3.

3.2. THE OBJECTIVE OF STUDY AND THE OBJECTIVE OF ANALYSIS

The evaluation of the IEMs in Table 3.1 concerning the objective of the study shows that many models are developed for analytical or methodological use. A reason that so much emphasis is placed upon methodological aspects may be that the development of IEMs is a young field of research which started in the early seventies. The methodological relevance of an IEM can be subdivided into:

(i)   Its relevance for the development of a mathematical tool to link phenomena from various modules. Examples from such developments in the survey are the compromise programming approach in an evaluation procedure (in Section 2.2), the use of input-output analysis in an optimization procedure (in Section 2.10), the use of simulation analysis in an optimization procedure (in Section 2.12), and the use of a scorecard evaluation procedure to display policy alternatives (in Section 2.5);

(ii) Its relevance for the methodological development of a model design to integrate phenomena from various modules. Examples of the progress in the development of the model design from IEMs can be illustrated by the multi-layer projection with the economic, employment and environmental modules represented as parallel layers (see Section 2.9), the modular form of the water management analysis for the Netherlands with a water distribution module which acts as the core of the analysis (in Section 2.4), and the modular form of an IEM applied to a regional housing development scenario which includes qualitative and quantitative analyses (in Section 2.14).

However, some models discussed in Chapter 2 have also been used in decision-making processes, and the first column in Table 3.1 shows that 7 models out of the survey have been developed for explicit use in regional decision making processes (e.g., regional development and fisheries, regional water management activities, environmental quality control).

The second column in Table 3.1 shows that the main objective of the analysis of IEMs out of the survey is subdivided into five categories, viz. data base generation, description, explanation, forecasting and evaluation. These five types of modelling objectives can be interpreted as follows (see also Burch et al., 1979; van Lierop,1986; Nijkamp, 1983).

The generation of a data base system means a structured organization of information in order to increase the insight regarding a certain phenomenon. The organization of information may be carried out from data base systems in various ways; for example, verification (validation of the correct nature of data), classification (or the grouping of information into specific classes), updating (interpolation or extrapolation when additional information becomes available, for example the RAS-techniques which are tools for updating of input-output tables), and retrieving (selection of specific information from specific media). An example of a data base generation from the survey was discussed in Section 2.13. A developed data base may be used for other types of modelling categories, such as description or exploratory analysis, and explanation etc. An example of an information and modelling system for environmental policy in the Netherlands is discussed by Hordijk et al. (1981).

Description or an exploratory analysis of phenomena is a structural representation of the information; for example, the state of the regional water flows in the Netherlands with regard to the PAWN study in Section 2.4, and the exploration of the spatial interdependencies of the water flows in the PAWN study.

Explanatory analysis or impact analysis deals with the stimulus-response relationships in causal models concerning the effects of public policy decision making processes. Model forecasting is the extrapolation of the stimulus-response relationships toward the future and it includes temporal components in an explanatory analysis.

An evaluation procedure copes with the process of analysing policy plans or projects, and it deals with the comparison of the advantages and disadvantages from various policy alternatives. Two types of evaluation procedures have been discussed in the survey, viz. a compromise programming procedure (in Section 2.2), and a scorecard evaluation of policy alternatives (in Sections 2.4 and 2.5).

3.3. THE MODULES AND THE MATHEMATICAL TOOLS

The third column in Table 3.1 shows the modules of the IEMs out of the survey
and the next column shows the mathematical tools which have been used to
operationalize the models. All IEMs discussed in Chapter 2 include at least
an economic module and a natural environmental module. Four types of mathematical tools are the most relevant ones in the survey, viz. optimization,
input-output analysis, simulation analysis and a scenario approach. Table 3.2
shows a cross-classification of the modules and the mathematical tools used
to operationalize these modules. The numbers in the table refer to the corresponding sections in Chapter 2. Such a table may indicate whether some
correspondence exists between modules and the mathematical tools used to
operationalize the modules.

Table 3.2. A cross-classification of modules and mathematical tools.

Mathematical	Module	
tool	Economic	Natural environment
Optimization	2.2; 2.7; 2.8; 2.9; 2.10; 2.12	2.2; 2.9; 2.10
Input-output	2.6; 2.7; 2.9; 2.10	2.6
Simulation	2.3; 2.11; 2.13; 2.15	2.3; 2.4; 2.5; 2.7; 2.8; 2.11; 2.12; 2.13; 2.14; 2.15
Scenario	2.4; 2.5; 2.14	

The cross-classification in Table 3.2 shows that optimization procedures and
input-output models have been used especially in case of economic modules.
Optimization models, which are mainly used in regional analyses, determine
the optimal allocation of public activities among regions, such as, the compromise programming evaluation between agricultural activities and nutrient
loading levels in water from Section 2.2. Input-output methods in regional
analyses are economic-based methods which originate from the double accounting principle in national accounts (see also Nijkamp et al., 1986 for a discussion of various modelling approaches in regional-economic analysis).
The results from Table 3.2 also show that the use of simulation models is a

widely used tool in modelling natural environmental processes. Simulation models 'have been derived from an a priori consideration of the essential relationships between the elements of populations or processes or from the integration and testing of the compatibility of information derived from a large number of different sources' (Jeffers, 1982, p. 23).

A number of well-founded reasons of the widespread use of simulation modelling in environmental processes were presented by Fedra (1983) when he mentioned 'the complexity and variability of environmental systems, the scarcity of appropriate observations and experiments, problems in the interpretation of empirical data, and the lack of a well established, comprehensive theoretical background' (Fedra, 1983, p. 19).

The scenario approach was only used in three models to operationalize the economic module. A scenario consists of a description of one or more future situations as well as of a description of the transformation process from the present state of a system to its state in the future. See for example the scenario approach which was used in the POLANO study in Section 2.5 where the consequences of restricted and unrestricted investment policies on recreational activities were analysed.

3.4. DIMENSIONS OF SPACE AND TIME IN INTEGRATED ENVIRONMENTAL MODELS

The evaluation of the IEMs in Table 3.1 concerning the spatial dimension shows that the IEMs are single regional models in eleven out of fourteen cases. See also Section 1.1 for a discussion of the primary efforts in regional-economic modelling, which were also developed at a single-regional scale.

The spatial dimensions as well as the temporal horizon may be different for the variables which describe the economic activities and the ecological processes (Nijkamp, 1984a). The spatial scale of ecological variables such as species diversity, is often represented in terms of square metres, and the economic variables may be adequately measured at a regional or sub-national level.

A reasonable example of the difference in time perspective and spatial scale in IEMs is the acid rain problem, with economic-technological solutions in specific areas which can at least in principle be implemented within a period of a decade. However, the observable environmental diffusion effects of such policy measures, which cross the boundary of regions and even of nations, may take several decades.

Clark (1985) presented some examples on the variation in time scales in case of climatic change between socio-economic and ecological processes. Socio-economic processes like crop cycles, industrial processes and population may cover a temporal focus between a month to a century, while ecological processes such as animal reproduction, and vegetation range extension may cover a temporal focus between a week and a period of 5000 years.

A region itself may also be defined in various ways such as the standard metropolitan labour areas (SMLA) in the United States which aims to define regional labour market areas, and the administrative regions (provinces, counties, etc.) which are the spatial entities used by regional authorities for the collection of demographic data. Some regionalization principles will be further elaborated in Section 4.5. External linkages may also exist for models operationalized at the single-regional or multi-regional level, because flows of goods, activities or money may exert their influence across the boundary of a spatially demarcated region (see also Nijkamp et al., 1986).

3.5. THE AVAILABILITY OF INFORMATION IN INTEGRATED ENVIRONMENTAL MODELS

The evaluation of IEMs in Table 3.1 showed that the lack of appropriate information was mentioned to lead to difficulties in seven cases out of the fourteen models from the survey to operationalize one or more modules. The difficulties mentioned by the modellers vary from the quality of information to the absence of information concerning modules. The quality of information may be poor when such information concerning variables is not known with precision.

Three points can be mentioned concerning the quality of appropriate information. First, the availability of statistical information at the regional level for economic based sectoral activities (production, employment or consumption, etc.) can still be improved (see also Klein and Glickman, 1977). The availability of information for the operationalization of an IEM depends on the definition and spatial demarcation of a region because the major part of socio-economic data is collected for administratively defined regions (e.g., states, counties, provinces). Data problems, such as the availability of statistical information, may lead to serious problems in case of estimation and validation of regional models.

Second, the evaluation in Table 3.1 shows that the lack of appropriate information becomes especially problematic for operationalization of ecological processes and activities. This varies from the poor quality of information to the missing of information for the ecological processes.

Finally, the models discussed in Sections 2.14 and 2.15 mention the various data sources and the different levels of measurement of variables within the framework of an IEM. The data may be obtained from different sources, varying from regional or national statistical offices, controlled experiments, surveys, panel studies, to expert judgments. The corresponding levels of measurement of the data vary from a metric to a non-metric scale. The various mathematical tools for dealing with such levels of measurement of information will be further elaborated in Chapter 5.

3.6. CONCLUDING REMARKS

The state-of-the-art of IEM modelling approaches was discussed in Chapter 2, to be followed by a confrontation of the models from the survey in Chapter 3 by making use of nine points of reference.

This introduction to integrated environmental modelling shows that IEMs may encompass among others the following elements: environmental impacts of production and consumption (in Sections 2.3, 2.4, 2.6, and 2.9), ecological processes in relation to the economy (in Sections 2.2, 2.5, 2.7, 2.11, and 2.12), resource management (in Sections 2.8, 2.10, and 2.13), land use impacts (in Section 2.14 and 2.15), spatial mobility and demographic elements (in Section 2.14) (see also Arntzen and Braat, 1983 for a discussion of the various types of application of IEMs). Such models aim to provide a comprehensive and systematic picture of the components and interactions in an economic and ecological module for a certain area.

The different ways of operationalizing economic, demographic and ecological phenomena concerning mathematical tools, space/time dimensions and the availability of information was discussed in this chapter. This difference shows the necessity to integrate such phenomena in a systematic way. Many conflicting issues have come about in recent years in regional-economic and environmental policy making, which is, for example, witnessed by the environmental impacts of acid rain in natural parks. Also many other cases (fishery, agriculture, recreational activities, e.g.) as mentioned in Chapter 2, demonstrated the need of a rigorous systematic approach to current planning.

A systems theoretic approach will therefore be discussed in the next chapter as a methodology to integrate phenomena which originate from various disciplines.

PART B:

METHODOLOGY AND TOOLS TO OPERATIONALIZE
AN INTEGRATED ENVIRONMENTAL MODEL

CHAPTER 4. A SYSTEMS APPROACH TO AN INTEGRATED ENVIRONMENTAL MODEL

4.1. INTRODUCTION

An IEM is characterized by a formal and coherent model structure for environmental phenomena, in which the constituents emerge from various modules whose nature originates from (spatial-) economics, ecology, demography, geography, transportation analysis, etc. The various IEMs presented in Chapter 2 showed that the phenomena and processes described in the modules differ in nature as well as in the contents of the variables. The economic part of an IEM may include, for example, such variables as production, consumption and employment, while the ecological part may include variables which reflect the association and diversity of ecosystems. However, despite the differences between these phenomena originating from various disciplines, they may interact with each other in an IEM. The examples of IEMs presented in Chapter 2, and followed in Chapter 3 by an overall discussion of the common characteristics, also showed the relevance of a systematic approach to the development of an IEM.

A sound research strategy is therefore necessary in order to structure, analyse and decompose the complex and mutually dependent environmental problems into relevant parts. Systems theory is such a research strategy having its roots in biology, psychology, philosophy, and economics. The development of a systems approach is influenced by the rise in complexity of society because 'an interrelationship exists between all elements and constituents of society. The essential factors in public problems, issues, policies and programs must always be considered and evaluated as interdependent components of a total system' (the Canadian Premier Manning in von Bertalanffy, 1968, p. 4). In this respect, <u>systems theory</u> can be regarded as a methodology towards the operationalization of a model with strongly interrelated subsystems. Ecosystems, for example, can be represented by means of processes and negative feedback relationships in a dynamic systems context. This argument is based on the consideration that an ecosystem is interpreted as a cybernetic system, with stimulus-response flows of information which are regulated by feedback relationships (see also von Bertalanffy, 1967).

The systems approach is based on two basic considerations, viz. (i) reality is considered from the behaviour of the 'whole' system, and (ii) a mutual dependence is considered between systems and their environment. Although the environment is outside the system, it may perform a permanent inflow or outflow of matter or energy to the system. A system having a mutual dependence with its environment is usually also called an open system.

The first part of this Chapter (Section 4.2) is devoted to a discussion of the systems approach in socio-economic and ecosystem studies. A brief historical overview, the main characteristics, and the usefulness of systems theory as a tool of integration in multidisciplinary research will be discussed first in that section. The relevance from various applications of systems analysis in ecosystem studies and in socio-economic studies will also be discussed in Section 4.2.

Systems theory will be used in the second part of this chapter (Section 4.3) in the framework of the development of an IEM. Two concepts of integration of an IEM are presented in Section 4.3, and a distinction will be made between a horizontal (or parallel) and a vertical (or hierarchical) model approach. The two approaches differ in the way that the links between modules are conceptualized and operationalized. A recently developed model concept of integration that partly reflects the characteristics of both the horizontal and the vertical model approach is the satellite model design. This concept of integration will be discussed in Section 4.4.

Integrated modelling efforts at a regional level are usually facing the problem of a relevant spatial demarcation. The IEMs discussed in Chapter 2 were developed at the regional level. However, the survey in Chapter 2 also showed that a region is not a uniquely defined spatial entity, and a region may also be subdivided into sub-regions in various ways. Ecosystems and economic phenomena, for example, may have spatially different effects which cross the boundary of administratively defined spatial systems, and such features may influence the model outcomes. Different ways of linking regions with each other or with other spatial entities will be presented in Section 4.5. A distinction will be made in that section between the so-called spatial aggregation level and the spatial scale level and their relevance for model outcome.

Some concluding remarks are presented in Section 4.6 with respect to a systems approach for developing an IEM.

4.2. SYSTEMS THEORY

The main developments of science since the time of Newton and Descartes were influenced by reduction of phenomena into isolated components. Phenomena and processes were reduced and decomposed into as many isolated and simple components as possible, while often the possible causal relationships between them were neglected. Within this view of reductionism, the aggregation of the individual components was considered to be an explanation of the phenomena themselves. As a consequence of this view, modern science is characterized by

an ever increasing specialization of disciplines each generating new subdis-
ciplines. However, a drawback of the increasing monodisciplinary developments
is the ignorance of the mutually dependent aspects which cross the boundaries
of a single discipline.

The neo-classical economic theory, for example, is based on the atomistic-
mechanistic world view; this world view is based on Newton's 'Philosophiae
Naturalis Principia Mathematica'. It is atomistic because the production
factors land, labour, and capital are considered to be separate components
during the production process. The production factors are only related to
each other by means of their relative values. The neo-classical approach is
also mechanistic because economic systems are considered to operate along
equilibrium values. Atomistic-mechanistic models are mainly charaterized by a
series of stable equilibria (see also Norgaard, 1985).

The development of systems theory is aimed at crossing the boundary of the
various scientific fields (or disciplines) and to provide a tool of communi-
cation between disciplines because 'one wonders sometimes if science will not
grind to a stop in an assemblage of walled-in hermits, each mumbling to him-
self words in a private language that only he can understand' (Boulding,
1956, p.12).

The development of systems theory has its historical roots in disciplines
like biology (Ludwig von Bertalanffy), psychology (development of Gestalt-
psychology which is aimed at clarifying human behaviour and phenomena by
means of 'Gestalts' or wholes), and economics (Kenneth Boulding). Von Bertalanffy mentioned that phenomena cannot be explained exclusively in terms of
its components, but should be analysed also in terms of the relationships
between the components. Conventional economics on the contrary, tends to
limit its field of operation to market oriented phenomena, mainly characterized by production and consumption activities. Boulding however, seems to be
a more problem-oriented economist; this means that he is convinced that there
is a considerable interdisciplinary area of social science and moral science
(e.g., ethics) where economists should be ready to learn from other disciplines.

A key element in the development of systems theory with respect to the various disciplines was the consideration that a system is defined as a unit
which consists of mutually related elements. This consideration is in agreement with the view on the world-order of the Greek philosopher Aristotle who
supported a composition law by stating that 'the whole is more than the sum
of its parts, and the part is more than a fraction of the whole', which still
holds true for the basic systems approach. An analysis of parts or components, which are isolated, cannot provide a total understanding of a functional whole.

Ludwig von Bertalanffy, who introduced and advocated the term 'General Systems Theory', published various statements concerning its aims in the first yearbook of the Society for General Systems Research in 1956, viz. (von Bertalanffy, 1956):

(i)   a general tendency towards integration in the natural and social sciences, because a consistent conceptual structure of physical, biological, and social phenomena appears to be necessary;

(ii)   the integration seems to be centered in a general theory of systems;

(iii)   the development of unifying principles of the individual sciences;

(iv)   a need for integration of scientific education.

A general system was in the early fifties defined as a theoretical system which is of interest to more than one discipline.

Systems theory in relation to global environmental problems has been used, for example, as the basis for the model upon which the first Report of the Club of Rome has been based. Forrester (1971) developed a simulation model concerning the major relationships between environmental aspects, growth of population, industrialization, natural resources and food problems. Such basic elements could not be analysed separately because of their mutual dependency. The mutual dependence between the different parts was emphasized in this Report.

The general economic equilibrium analysis developed by Walras and operationalized by Leontief in the input-output analysis is an example of a general systems approach in economics. Such equilibrium analysis in economics can be generalized toward other fields like ecological equilibrium analysis, which is based on a number of interacting populations.

A hierarchical ordering of phenomena is a fundamental concept in systems theory. An example of this hierarchy in nature are atoms, which combine to form with their own characteristics, and which can be combined to biological organisms. The organisms give phenomena which in turn lead to even higher levels of organization (von Bertalanffy, 1968; Rahmatian, 1985; Simon, 1973). Simon makes a distinction between physical, chemical, biological, social and artificial systems, which are denoted as hierarchical systems, and are linked to each other.

Formally, a system is a collection of objects, components or modules with interactions or input/output statements between them (see also Mesarovic, 1964; Wymore, 1976). The objects may also be called entities. The mutual dependence and hierarchical ordering of objects is a major point of systems theory, as was already mentioned by von Bertalanffy and Simon.

An example of a closed system is presented in Figure 4.1. The example will be called a closed system, because no input effects from outside the system are considered. A closed system is isolated from its environment, because there is no exchange of matter, energy or information.

Figure 4.1 consists of components A and B which are mutually dependent. The output of component B becomes input for component A and vice versa.

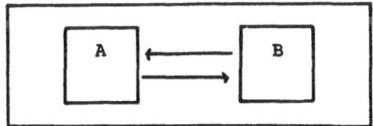

Figure 4.1. Schematic representation of a closed system.

The structural properties of such a system can be summarized in the following way (see Bennett and Chorley, 1978):
(i)   the properties of each component will affect the system as a whole;
(ii) the properties of each component will affect at least one other compo-
      nent.
A system like the example in Figure 4.1 cannot be partitioned into sets of mutually independent components. This is in agreement with the history of systems theory: a system has properties which are quite different from the separate components, and systems theory is mainly concerned with 'complicated systems where components exhibit a high degree of interdependence. The behaviour of the whole system is then usually much richer than the sum of the parts' (Wilson, 1981, p. 3).
A closed system like Figure 4.1 is isolated from its environment by a boundary. An example of a closed system is a chemical equilibrium which may be established when a number of reactants are brought together in a vessel. Another example of a closed system in a mathematical form is a simultaneous set of equations with endogenous variables, but no exogenous variables (or 'constant level' variables from outside the system).
Boulding, for example, considered a circular course process in economics, with input and output factors which are related to each other by so-called 'throughputs'. Resources are inputs to the production and consumption process and are transformed into output. The main emphasis of economics is placed upon the process of 'throughput'. Because of the finite capacity of resources and energy, the open system process of input-throughput-output has to be changed in a closed system marked by a circular course with output to be used again as input factor. Because of these characteristics he compared the nature of an economic system with a spaceship (his notion of a so-called 'spaceship earth') with energy in the transformation from input to output,

which acts as the limiting factor in the production process.

The circular course process which Boulding considered in economics has been elaborated further in ecological systems by means of <u>materials balance models</u>. Such models are closely related to three characteristics from thermodynamics, viz. the conservation of matter, the conservation of energy (the first law of thermodynamics), and the increase of entropy for a closed system (the second law of thermodynamics). The input of materials balance models consists of flows and stocks of energy and materials which are transformed into final goods and residuals (see also Kneese et al., 1970; Eriksson, 1984). Some limitations of the materials balance models are:

- its physical basis which precludes an appropriate analysis of psycho-somatic impacts of specific pollutants (toxic chemical compounds, e.g.);
- ecological processes which are in general neglected;
- various economic aspects which cannot be dealt with (the monetary part, the societal processes, and so forth).

Phenomena which do not belong to a defined system will be called its environment. Input effects from outside a system may also be possible. Such a system will be open in nature (see the example in Figure 4.2). Every living organism may be interpreted as an open system with a permanent inflow and outflow of components.

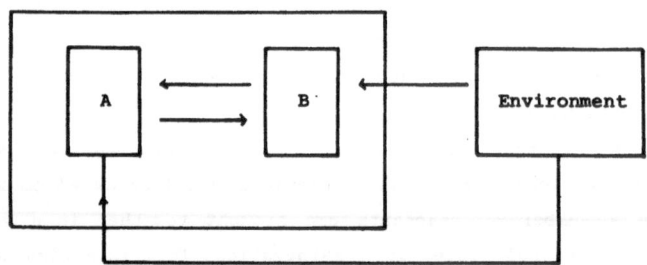

Figure 4.2. Example of an open system.

A closed system simply consists of an input, say A, and an output, say B, which are linked to each other in an analytical way by a transfer function f, or in formal terms by means of B=f(A). The transfer function f which links two components A and B is the main element in the system, since it describes the process or changes from systems input to systems output. Such simple systems where an input is transferred into an output by means of an operator or transfer function are also called black box systems. Feedbacks between system components are an essential part of systems analysis, and the basic elements of feedback relationships will be presented below in terms of a closed system. A feedback relationship exists, for example, between A and B in Figure 4.1 because the output of component (or sub-system) A, which is influenced by the output of sub-system B, again becomes input for sub-system B. The stimulus-response relationships between sub-systems may be denoted by a so called correlative system (Jones, 1983). An example of a correlative system for electrical energy demand is given in Figure 4.3 (see also Casti, 1979).

A correlative system only presents the links between key factors, which may be included in a mathematical representation of a system as in Figure 4.3. The lines which link the components are lines of a probable relationship.

A feedback relationship in Figure 4.3 exists from component 7 back to component 4. The information concerning the effects of a process which is the output from component 7 is returning to one of its sources in component 4. The feedback mechanism as mentioned in Figure 4.3 is the basic characteristic

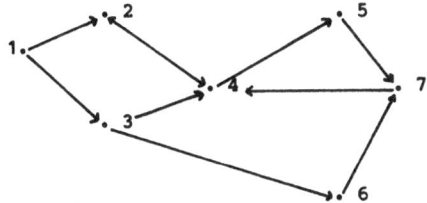

with: 1 = energy capacity   4 = energy use
       2 = energy price   5 = environmental quality
       3 = number of factories   6 = number of jobs
                                    7 = population

Figure 4.3. Correlative system of electrical energy demand.

of cybernetics and is aimed to control a system's function, either to maintain a desired state or to reach a specified goal.

It is worth noting that in various cases systems theory has become a rather confusing term, as it is often claimed by many researchers that they follow a systems approach, without precisely defining the system at hand. A variety of systems approaches have been developed in case of, among others, problem-solving and decision-making, planning, or of describing and explaining systems and their behaviour (see also Rahmatian, 1985 for a presentation of applications on systems approaches). Some general characteristic features of systems theory will therefore be mentioned now:

(i)   information concerning systems is obtained from different sources and
       disciplines;

(ii)   a system is based on a concrete real-world situation, and systems theory is hence an experimental science which includes human activities, especially those which occur at specific locations and times;

(iii)   systems theory is usually focussing attention on decision making (e.g., in regional planning) with complex multidimensional problems;

(iv)   systems theory provides a unified framework of analysis, so that the use of mathematical-statistical tools and of computers is usually appropriate.

The major tasks of a systems analyst are therefore twofold:

(a)   the design of a coherent system that can be used to analyse the scope of a problem under consideration. The relevance of a system cannot be proved in a mathematical way, but should be based on experiments (see (ii) above);

(b)   the design of a set of formal (mathematical statistical) representations or models in order to analyse a real world problem in an adequate way.

A mathematical model provides a systematic structure for placing the information from a systems approach into proper perspective. The different stages of systems theory to relate a system with a mathematical model representation are summarized in Figure 4.4 (see also Beck, 1979; Bennett and Chorley, 1978; Caswell et al., 1972; Jørgensen, 1983; Morgan, 1981; Murphy, 1985; Svedin,

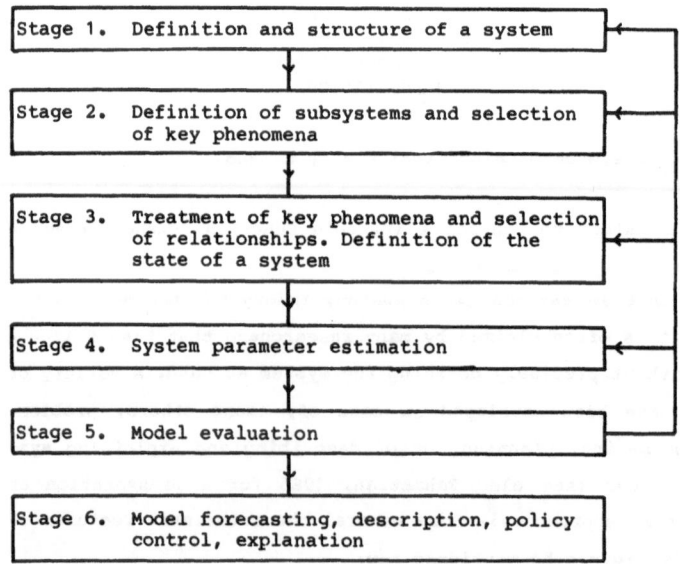

Figure 4.4. Definition and evaluation of a systems model.

1985; Wilson, 1981).

The six stages developing a systems model in Figure 4.4 can be explained in the following way:

Stage 1. A research question should be formulated which also defines the goal of the analysis. The first step in the systems process which is evidently influenced by subjective choices concerns a partitioning of the observed universe into a system and its environment. Preliminary choices should be made among others in relation to the temporal, spatial and sectoral scale within which the activities take place.

The time units to be used may vary from a long-term horizon of about ten years for dynamic physical planning systems to a short-term time horizon such as months for ecological systems.

The system to be analysed may be based on the spatial scale of a region, but also - in other cases - on a much more refined spatial scale with grids of 100 m^2.

Finally, the sectoral elements represent the number of components, i.e. either aggregated data or more refined disaggregated data when a subdivision will be made according to, for example, types of labour or economic activities (agriculture, industry or service sector, e.g.).

The requirements to specify a systems model in case of socio-economic and environmental research have been further elaborated by, among others, van Lierop (1986) and Sommermeyer (1967).

Stage 2. This stage is based on the selection of necessary and useful objects, behavioral features and state variables. Necessary aspects to be considered are:

(i) the relationship between the variables in a system (which depends on the aim of the study);

(ii) the inter-relationship of the system with sub-systems and the environment.

Stage 3. The state of a system implies a characterization of a set of relevant phenomena and a choice of the transfer function by means of specified values for the system. Variables which are not observable are called latent variables (for example regional welfare, environmental quality). In case of latent variables, manifest variables should be included to approximate the unobservable, latent variables, while in this context also latent variables techniques (like LISREL) may be applied. The model specification depends on the results from stages 1 and 2 as well as the availability of information.

Stage 4. In this stage the numerical significance of systems relationships will be operationalized and estimated. Behavioral features can be denoted in a qualitative or a quantitative way.

Stage 5. The level of explanation and the fit of the model can be checked by means of statistical test procedures, based on the efficiency and fit properties from the parameter estimates. The previous stages of the analysis may be reformulated or adjusted when the parameter estimates are inefficient or suboptimal in some respect due to, for example, low significance level, spatial or temporal autocorrelation, etc.

Stage 6. The fitted model may be used for description, planning, policy and forecasting; this use depends on the goal of the analysis in stage 1 and the results from all previous stages.

The different stages of systems theory in Figure 4.4 show that a synthesis between information collection and mathematical modelling can be achieved in five successive stages of system development (Jeffers, 1976, and Nadler, 1985), viz.:

(i)   definition of objectives and preliminary analysis in terms of the hierarchical nature of phenomena;

(ii)   experimentation and collection of information;

(iii) management and expert judgements by decision-makers and modellers;

(iv)   evaluation of the results achieved;

(v)   final synthesis between the proposed systems design, the information obtained, and a mathematical model operationalized.

The six stages of Figure 4.4 to extract information for decision making from the integration of socio-economic data and natural resources data is termed a geographic information system by De Man (1984).

The basic principle to identify ecosystems' behaviour is the consideration that the pattern of interactions between the phenomena or objects is at least as important as the characteristics of the phenomena themselves (Caswell et al., 1972), because of the feedback relationships between processes.

Ecological processes which are multidimensional in nature, may involve a number of uncertainties and risks, its phenomena may be nonstationary (or time dependent), and nonlinear in nature (see Bennett and Chorley, 1978).

Uncertainty of ecological processes deals with the scarcity of the available data for the analysis which are also subject to sampling errors (see also Fedra et al., 1981 for an example of ecosystems modelling with uncertainty).

Having discussed now the main characteristics of systems theory in case of modelling economic and environmental phenomena, the systems approach will be

presented in the next section as a methodology to develop the design of an
IEM.

4.3. TWO DESIGNS OF INTEGRATION

4.3.1. Introduction

The first stage in the process of the definition and evaluation of a systems
model deals with the definition and structure of a system, and the second
stage deals with the determination and the definition of stimulus-response
relationships between sub-systems (conform Figure 4.4). An overview of systems approaches in the framework of analysing environmental phenomena is
presented by Gottinger (1974). The phase of model design to the operationalization of an IEM is discussed in this section and it defines recognition,
definition and structuring of a problem as well as the process of hypothesis
building concerning the formulation of a model in terms of its stimulus-response relationships (Müller, 1981, 1983). Two designs of integration concerning IEMs will be discussed, viz. a so-called horizontal or a parallel
and a vertical or hierarchical model design of integration. Nijkamp (1984)
indicated some conditions for a meaningful integration of economic and ecological models, viz.:
1. The models should provide a coherent and consistent representation of a
   system with economic and ecological components.
2. The analysis should have a satisfactory degree of disaggregation and variation, so that the differences between various sub-systems become appar-
   ent.
3. The interactions within and between the sub-systems should receive sufficient attention so that relevant links are described in a satisfactory
   manner.
4. An empirical application of the analysis should, in principle, be possible
   (either in quantitative measures or in a qualitative measure).
5. The analysis should provide sufficient and relevant information for a
   policy-maker who has to judge the various alternatives.
In the light of these considerations the following design principles may be
mentioned for integrated modelling efforts (Brouwer et al., 1984):
- relevance: which problem requires which type of integration?
- operationality: which approach assures a maximum of empirical or policy
  use?
- specification: which model structure leads to an ordering of multidisciplinary phenomena which is in agreement with a complex multidimensional

reality?

- <u>flexibility</u>: which model type is suitable for new alternative applications and extensions?
- <u>accessibility</u>: which model structure can easily be employed by other users or at least be made transferable to analogous problems?
- <u>policy transparency</u>: which model type provides a clear insight into relevant alternative choice options, and in policy effects?

The next subsection will be devoted to a discussion of a horizontal and a vertical model approach. Both approaches have been used in empirical applications. An example of an horizontal model structure and a vertical model structure of IEMs discussed in Chapter 2 will also be presented.

A confrontation of two empirical applications of the horizontal and the vertical model approach makes use of the six design principles.

4.3.2. <u>A horizontal model approach and a vertical model approach</u>

The first stage in the definition of a systems model in Figure 4.4 deals with the definition and structure of a system, also called the systems design phase. The model design for IEMs can be subdivided into a <u>horizontal model approach</u> and a <u>vertical model approach</u>. The distinction in these two approaches may indicate that the relationships between variables from different modules can be analysed in different ways (Brouwer and Nijkamp, 1986d).

A <u>horizontal model approach</u> is characterized by its interactions between monodisciplinary modules in terms of a parallel design. All relevant modules have an equal contribution to the model conceptualization and operationalization and will be analysed in equal detail. No clear distinction will be made in the model design between the impacts between variables from one module on the one hand and the impacts between variables from various modules on the other hand.

An example of the horizontal model approach is the multiobjective decision making analysis for the reduction of phosphorus loading in water, discussed in Section 2.2. The objective function consists of the sum of two dimensionless figures (see equation 2.1) which originally were measured in terms of the percentage reduction of phosphorus loading and yields of agricultural activities (see Figure 2.1). An advantage of the horizontal approach used in the example is that variables which originate from different disciplines (economics and ecology) are linked to each other in a rather straightforward way in this model by means of the multi-objective framework. The links between modules in a horizontal model approach of this example can be interpreted as the interactions between dimensionless figures which originate from an economic module and a natural environment module.

A vertical model approach is characterized by a hierarchy of the modules in a model with one or more of the modules being superior to all others. The vertical model approach places special emphasis on the relationship between the dominant module and the other modules. The hierarchy of modules will be relevant in empirical situations when the level of one factor is the major point of study which is an input for all other modules. However, the mutual relationships - except the ones between the dominant module and the other modules - are considered to be of minor importance in the vertical model approach.

An advantage of the hierarchical nature of a systems model design is the identification of 'the interrelationship among levels and subsystems' (Nadler, 1985, p. 690).

A vertical model approach, distinct from a horizontal model approach, is mainly selected in empirical situations because 'initial attempts to cover all topics in a similar degree of detail have proved to be overambitious, in terms of staff time and data availability, and more recently it has become almost standard practice to adopt an approach focusing upon selected topics with major implications for policy or for short-term investment programs' (Batey, 1984, p. 65). The remaining possible relationships receive less attention and are assumed to be of only minor importance.

The characteristics of the vertical or hierarchical structure have some advantages when compared with the horizontal model structure (see Goodall, 1976) because it gives special emphasis on the major relevant links between modules. Such emphasis on the major relevant links between the modules may give an easier and better understanding of the impacts in a system. A practical advantage of the hierarchical approach may deal with data availability and staff time.

The hierarchical model approach corresponds to a representation of biological activities, for example, in a hierarchical way. A hierarchical systems approach in biology was developed by Rosen (1972), and he defined two conditions for an hierarchical systems organization, viz. '(i) the system is engaged simultaneously in a variety of separate distinguishable activities, and (ii) different system descriptions are necessary to describe these several activities' (Rosen, 1972, p. 59).

An example of the vertical or hierarchical model structure has been discussed in Section 2.4 for the analysis of availability, distribution and transport from water resources in the Netherlands (see also Figure 2.3). The central part of the analysis is made up by the water distribution module; a balance between supply of and demand for water from the different sectors (e.g., industry, agriculture, lakes, treatment plants) is obtained from and regula-

ted by that core module. The hierarchical model structure is in this example
based upon two levels, viz. the core level with the water distribution module, and the. lower level, which is defined by the core module, and which
consists of the other modules.

The advantage of the vertical model approach in Section 2.4 is the selection
of one module as the dominant one. The different aspects related to the distribution of water, such as agriculture, environment and industry are treated
in the other modules and analysed independently from each other. However, a
disadvantage of the vertical approach is that no direct links are modeled
between the other modules. Such lower level modules are only related indirectly to each other in the frame of the water distribution module.

An example of hierarchical decomposition of socio-economic systems is presented by Müller (1981; 1983) for the Melle area in West Germany. A socio-
economic system has been decomposed into a subsystem, and a module has been
specified for each cluster of variables.

A confrontation of the horizontal model approach and the vertical model approach concerning the principles of relevance of integration, operationality,
specification, flexibility, accessibility and policy transparency is summarized in Table 4.1.

Table 4.1. Comparison of two model concepts for IEMs on several criteria.

relevance	A horizontal approach may be relevant in a monodisciplinary evaluation procedure of economic and environmental phenomena. A vertical approach may be relevant when one module is dominating the other modules (water distribution in watermanagement, e.g.).
operationality	Both approaches have been operationalized in an empirical application.
specification	A horizontal model approach has a stronger tendency to specify modules along disciplinary lines than the vertical model approach, because the latter one emphasizes an analysis of key processes and factors.
flexibility	A disadvantage of the vertical approach may be that this type of model may have to be entirely redesigned, and extended when new applications are needed.
accessibility	A horizontal framework may contain a lot of information that is irrelevant to the (outside) user.
policy trans-parency	A clear insight into relevant alternative choice options may be obtained for each of the two model concepts. Policy transparency can be interpreted as the result of the evaluation from the five other mentioned points of confrontation.

4.4. TOWARDS A SATELLITE DESIGN OF AN INTEGRATED ENVIRONMENTAL MODEL

The conclusion from Section 4.3 concerning the relevance of the horizontal model approach is that it may become rather overambitious and complicated, in terms of design, staff time, budget and data availability, to operationalize such a model. This becomes especially severe when various modules such as demography, recreation, transportation and built environment are included in an integrated regional environmental modelling approach.

An advantage of the vertical model approach is clearly its hierarchical structure with one (or more) modules being mutually dependent on all these other modules, and the presence of only a few factors which are input for all these other modules. However, a disadvantage of the vertical model approach is that no direct links are considered between the lower level modules. Such lower level modules are only related to each other in an indirect way by means of the higher level module(s).

Table 4.1 showed that both model designs appear to have advantages and disadvantages.

A mixture of the horizontal and the vertical model design will be introduced in this section, and this concept of integration for an IEM will be called the satellite model design.

There are three reasons to develop a satellite design for IEMs, viz.:

a) a systematic approach is necessary because of the complex nature of IEMs, with phenomena and processes which originate from different disciplines and are to be analysed in an integrated way. Systems analysis can be regarded as a methodology to unify the principles of the individual sciences, with subsystems that cannot be partitioned into mutually independent components;

b) a hierarchical nature may be useful when one or more modules are the key factors for the analysis and are also an input for the other modules;

c) a distinction may be desirable between monodisciplinary relationships and multidisciplinary relationships which cross the boundaries from single disciplines.

The satellite principle can be characterized by three features, viz. (see Brouwer and Nijkamp, 1986d):

(i)   One module is considered as the central part of the analysis (i.e., the core of the model), based on a priori knowledge or a priori assumptions concerning the impacts between variables. The main focus of the analysis is oriented toward this module. All characteristics of this core module which are related to the other modules are determined in the first step;

(ii) All central aspects which originate from the first step are linked to each other, and they form the core module of the analysis;

(iii) All other aspects of the analysis which do not refer to the core module, are finally linked to each other in the third step. Direct relationships may also exist between the modules which do not belong to the core module, and such interrelationships indicate the level of integration of an IEM.

The satellite model design shares characteristics of both the horizontal and the vertical model approach. The hierarchical nature of the satellite model design in steps (i) and (ii) is the main characteristic of the vertical model principle. The links between variables from different modules in step (iii) denote the main characteristic of a horizontal model approach.

Because of the three characteristics of the satellite model design, three types of relationship exist for such a model principle:

(a) relationships between variables within a module, also called intra-module relationships;

(b) relationships between modules, also called inter-module relationships, which are at a more aggregated level than those of (a) and usually multidisciplinary in nature;

(c) relationships focussing on the core of the analysis which are monodisciplinary in nature.

An example of the satellite principle will be presented below, which is a conceptual representation of the water distribution analysis discussed in Section 2.4. The water distribution module is the core of the analysis and the main links of the model are summarized in Figure 4.5.

All phenomena in Figure 4.5 have direct relationships with the distribution and transportation of water in the area. The environmental aspects, for example, include the analysis of water quality which may be denoted in terms of chloride concentrations, eutrophication or BOD-levels.

In the satellite concept of integration for such a model, the first step includes the water distributional aspects which are related to the various, separate modules. Such relationships are monodisciplinary in nature. The water distribution module is the core module and regulates distribution and transportation of water.

The hierarchical model structure within the satellite concept of integration, such as the example in Figure 4.5, is based upon three levels, viz.:

(i)   the core level, which represents the water distribution module;

(ii)   the second level, which consists of the aspects from the core module, related to the other modules (environment, IJssel lakes, drinking-water companies, and agriculture);

(iii) the third level, which consists of the lower level modules and their interrelationships.

This section showed that the satellite principle is a useful approach for the structure of a cross-disciplinary model. It is based on a modular design of variables and structures of the model concerned, and implies that the kernel of an IEM is made up by the key mechanism of a certain economic-environmental problem area. Having identified the key mechanism of a cross-disciplinary model, all other (intra-, inter- or multidisciplinary) components can be added as nested derivatives of processes taking place in the core.

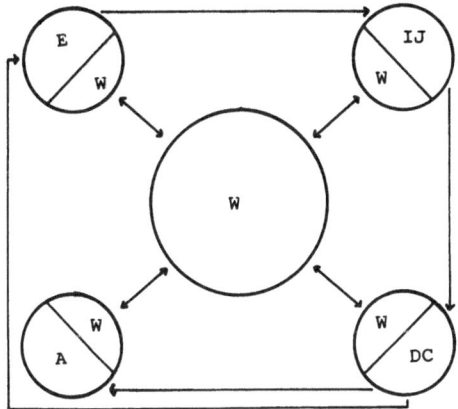

W = water distribution
IJ = IJssel lakes
DC = drinking-water companies
A = agriculture
E = environment

Figure 4.5. A correlative representation of an IEM with the satellite principle.

4.5. THE SPATIAL AGGREGATION LEVEL AND SPATIAL SCALE LEVEL IN REGIONAL ENVIRONMENTAL MODELLING

Spatial aspects of an IEM have already been touched upon in previous sections. All models discussed in Chapter 2 have a spatial coverage at the regional level. Some of them are also multiregional models because they show a subdivision into sub-regions which may be linked to each other. The relevance of the spatial aspects of the model results of IEMs at the regional level will now be discussed.

In the sixties the spatial aspects in geography needed the use of mathematical theory to be operationalized as a scientific classification procedure.
Various classification and clustering techniques have been developed since
that time (see also Fischer, 1980). A general definition of three basic components in geography - a set of places, a distance metric and a measure of
area - was recently developed by Beguin and Thisse (1979) by making use of
measure theoretic characteristics. However, in empirical applications of
regional integrated environmental models a serious disadvantage of the mathematical partitioning of space is that it does not take into account 'human
behavioural processes such as perception, learning, attitude formation and
decision making' (Gale and Golledge, 1982, p. 60).

In general, a classification procedure for a regional system may be based on
four principles (see also Paelinck and Nijkamp, 1976; Rietveld, 1984), viz.:

(i)   administrative regions, which are the natural spatial unit for the cor-
      responding regional authority, and are classified from an administra-
      tive or political point of view (for example, the partitioning of the
      Netherlands into 40 so called COROP regions, or into 128 labour market
      areas);

(ii) homogeneous regions (also called natural or ecological regions), class-
      ified by means of any selected homogeneity principle, with a high de-
      gree of uniformity or correspondence concerning a chosen attribute (for
      example, the degree of urbanization, the population density, or the per
      capita income);

(iii) functional regions, which indicate a high degree of interdependence
      concerning two or more attributes. A functional region may, for exam-
      ple, be defined by the interaction of goods, services, capital and
      labour such as the standard metropolitan labour area (SMLA) for region-
      al labour markets;

(iv) nodal regions (or core regions) in which the internal economic rela-
      tionships are more intensive than the relationships with respect to
      regions outside the area.

A region can be defined as 'a set of spatial points that are either homogeneous with respect to some characterization (criterion of homogeneity) or are
more intensively interrelated among each other than with other spatial points
(criterion of functional dependence)' (Siebert, 1985, p. 126).

The administrative regions have well-defined boundaries according to historical or political phenomena, while most of the homogeneous regions do not
have well-defined boundaries. A homogeneous region may, for example, be a watershed, as used in natural resource inventories.

In a systems theoretic approach to spatial aspects of an IEM, two classification procedures are applied, viz. homogeneous regions determined by attributes in regions, and a functional regional classification determined by multiple relationships existing between regions (see also Fischer, 1978).

A homogeneous or natural region may be characterized by, for example, geographical, physical, climatological and economic factors, while a functional region may be determined by a binary relationship between attributes and their spatial units. A binary relationship indicates whether or not some characteristic (for instance, migration) exists between a pair of regions, which can be represented by graph-theoretical tools.

A region can therefore be defined by means of a variety of characteristics. Various types of models can also be used to represent the linkage structure between regions and the national level, for instance, in linking regions in integrated environmental-economic models. A distinction can then be made between four types of models when the spatial structure of (multi)regional models will be analysed with the links between regional and national elements as the main point of emphasis (see also Curry, 1983 for a spatial statistical systems approach; Klein and Glickman, 1977 for a discussion of econometric model building at the regional level, and Nijkamp and Rietveld, 1982 for an overview of various types of regional modelling efforts). These types of models are:

a)   top-down models;

b)   bottom-up models;

c)   interregional models;

d)   (inter)regional-national models.

An example of each type of linkage structure based on three regions and the country level is presented in Figure 4.6.

The four types of multiregional models are classified by <u>functional relationships</u> between spatial units and by relationships such as interactions of goods, services, capital, labour or information (Fischer, 1983). In a <u>top-down</u> approach, the national variables are the key-factors for the regional variables without allowing effects from regional variables on the national variables. A multi-regional model can be derived immediately from a national model in that case. An application of a multiregional model with a top-down approach is developed by Lundqvist (1981) for Swedish regional planning, and denotes the linkages between economic planning, energy planning and regional development planning. However, in a <u>bottom-up</u> approach, the links between regional and national variables are just the other way round, because national variables can be derived from regional variables, e.g. national variables aggregated from regional variables. An <u>interregional</u> model is mainly characterized by its links between regions, when the interregional relations are

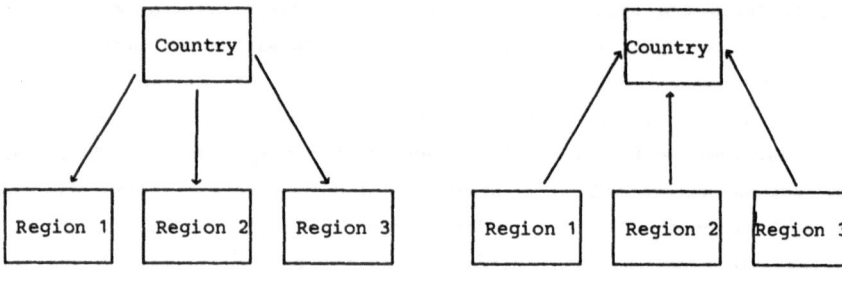

Figure 4.6(a). Top-down model.   Figure 4.6(b). Bottom-up model.

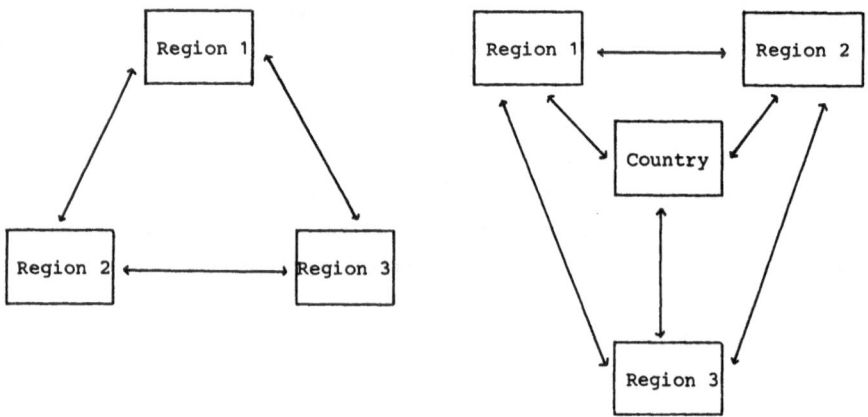

Figure 4.6(c). Interregional model.   Figure 4.6 (d) Interregional-national
                                          model.

Figure 4.6. Linkage structure of multiregional models.

not limited by national variables. A combination of these model structures,
with a simultaneous determination of the regional and national variables, is
called an interregional-national model.

The validity of nearly all applications of quantitative techniques with spatial data depend on the assumption that the spatial units are considered to
be given a priori (see also Coombes et al., 1982 for a discussion of functional regions for the population census in Great Britain). This approach has
often been questioned by regional modellers as a satisfactory geographical
assumption for regional applications. The relevance of zoning systems for
defining regions and sub-regions will therefore be discussed first.

It is easily seen that there is a large number of different ways by which any
study area can be subdivided and modified into non-overlapping areal units.

Consider for example Figure 4.7 as a spatial entity with respectively eight, four and two sub-regions. The selection of only one of the different ways of partitioning an area into subareas is also termed the Modifiable Areal Unit Problem. The relevance of the zoning system for the interpretation of modelling results was already mentioned by the statisticians Kendall and Yule in 1950 when they wrote that 'our correlations will accordingly measure the relationship between the variates for the specified units chosen for the work. They have no absolute validity independent of these units, but are relative to them. They measure, as it were, not only the variations of the quantities under consideration, but the properties of the unit-mesh which we have imposed on the system in order to measure it' (Kendall and Yule, 1950, p. 312). This means that the significance of statistical measures like correlation coefficients depend on the meaningfulness of the spatial units on which such measures are based, analogous to the dimensions in which variables are specified.

The zoning system for defining regions is based on two separate but closely related points, viz. the selection of the scale level and the selection of the aggregation level (see also Openshaw and Taylor; 1979; 1981).

The scale level is the variation in model results that may be obtained when increasingly larger spatial units of analysis will be selected for a set of areal data. However, there are also many possibilities of partitioning an area into a fixed number of spatial units, viz. when the scale level is defined. This is called the aggregation problem which will also affect the model results. The choice of both the spatial scale level and/or aggregation level determines the model outcomes. The aggregation problem is solved when a classification of subregions is determined by means of functional regions. The functional relations refer to multiple relationships between regions with a number of attributes which become a functional unit. In such a case the scale level still affects the model results (see Fischer, 1978).

The information obtained from a census, for instance, may be aggregated into local, regional or national level. The different zoning systems, viz. selection of the scale level and the aggregation level, provide alternative results when quantitative measures (parameter estimates, mean values of variables, correlation coefficients) are determined for each zone. The choice of the scale level is closely related to the different modelling results for applying the micro and macro approaches in economic analysis and the ecological fallacy problem in sociology (see also Langbeim and Lichtman, 1978). An ecological fallacy occurs when modelling results based on aggregated data are assumed to hold also for individual observations from the sample which form the zones being studied.

Openshaw proposed a geographically oriented algorithm of the above mentioned Modifiable Area Unit Problem to determine regions/sub-regions. The algorithm he suggested is called the Automatic Zoning Procedure and consists of a series of steps (see also Openshaw, 1983):

(i) selection of the scale level, viz. the number of regions required in the analysis;

(ii) selection of an objective function which will be optimized (for example, the correlation coefficient between variables);

(iii) a level of aggregation will be computed in an iterative procedure by optimization of the objective function.

In the context of the spatial scale of models, Openshaw and Taylor (1981) have presented an example about the wide range of possible correlation coefficients for different scale levels and aggregation levels in the 99 counties of the State of Ohio. The two variables considered are the percentage of the county population aged 60 and over, and the percentage of the county vote going to the Republican presidential candidate. The correlation coefficients they presented for the two variables and their data set had values in the range between -0.999 and 0.999 when the spatial units are aggregated to 6 zones. They claimed in their applications that 'a million or so correlation coefficients' could be obtained by changes in the scale level and the aggregation level.

A small illustration of the consequence of the scale level and aggregation level for correlation coefficients is given in Figure 4.7. This figure shows a spatial entity which has been subdivided into four types of non-overlapping areal units. The spatial entity has eight, four and two sub-regions, respectively. The numbers in this figure denote nominal values of two variables, one of them between brackets.

The correlation coefficients (r) in Figure 4.7 with values in the range between 0.00 and 0.94 depend on the chosen scale level and aggregation level. The effect of the different scale levels is represented by analyses (1) to (3) in this figure with 8, 4 and 2 regions respectively. The difference in correlation coefficients for two levels of aggregation is given in (3) and (4) from Figure 4.7.

Other empirical examples of the sensitivity of model outcomes in regional-economic applications for the scale level and aggregation level are given by Lohmoeller et al. (1985), on the causal influence of unemployment on votes for the NSDAP in Germany during the thirties for different regional aggregations, and by Nijkamp et al. (1984), on the determinants of the regional distribution of the number of persons receiving disability allowances in the Netherlands, with the scale level varying from a county, via a province to a

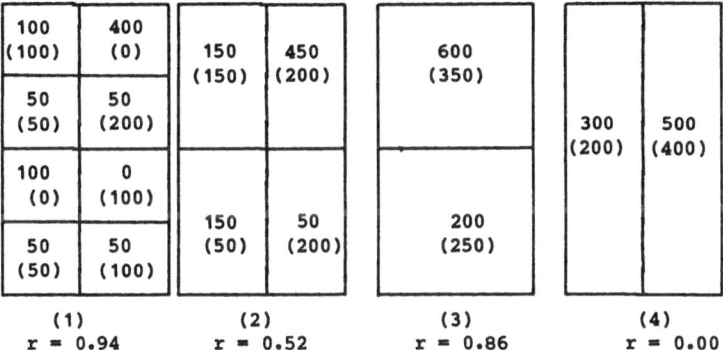

Figure 4.7. Illustrative example of scale level and aggregation level.

national level.

This section showed that the information obtained for planning, analytic or evaluation issues may be highly dependent of the spatial aggregation level and the spatial scale level.

4.6. CONCLUDING REMARKS

A methodological framework has been presented in this chapter to analyse among others, environmental, economic and demographic phenomena in an integrated way. Systems theory has shown to be a useful approach to the development of an IEM.

A recently developed design of integration in a systems theoretic framework is the satellite design which includes the advantages of the horizontal and the vertical model design of integration, because (i) it includes a hierarchical nature of a model with one or more modules which are the key factors for the analysis, and (ii) it includes the direct relationships between the modules which do not belong to the core module. Such a satellite design of integration is interpreted in three levels, with higher order levels which determine the nature of lower levels. The various levels within the satellite concept consist of the modules.

The advantages of a systems approach in the design phase and operationalization phase of an IEM within a satellite concept of integration are:

- the modular design of variables and model structure of an IEM which is analogous to a systems framework with sub-systems;
- the hierarchical representation of modules representing the key phenomena in terms of three levels;
- the cross-disciplinary relationships of an IEM which are analogous to the

integrative nature of sub-systems in a general theory of systems.
The design and framework of integration to operationalize an IEM has been
presented up to now with a systems approach as a methodological tool. Some
mathematical and econometric tools which are relevant to operationalize an
IEM will be discussed in the next chapter.

CHAPTER 5. STATISTICAL AND ECONOMETRIC TOOLS TO OPERATIONALIZE AN INTEGRATED ENVIRONMENTAL MODEL

5.1. INTRODUCTION

A systematic approach to the design and framework of integration of an IEM was presented in Chapter 4 in terms of systems analysis. A system was assumed to consist of a set of interdependent elements which form a unified whole. A satellite concept of integration was then discussed in Section 4.4 in order to operationalize the mutual dependence between various modules within the framework of a systems approach.

However, a wide variety of aspects still remain for the operationalization of an IEM. Their specific relevance will be discussed in this chapter within the framework of integrated environmental studies (see also Brouwer et al., 1985 for a presentation of issues in IEMs with a systems approach). One of the conclusions from the discussion of the IEMs in Chapter 2 and their evaluation in Chapter 3 concerned the availability of information: information is often not known with sufficient precision for some models, either because the data are of poor quality, or because no sampled data are available for the impacts between variables. This situation holds particularly true for environmental models (see also Fedra, 1983 for the modelling of environmental systems under uncertainty). A set of mathematical tools are therefore discussed in this chapter that deal with binary or qualitative approaches concerning the analysis of the impacts between variables, as well as some tools dealing with the analysis of models which include variables measured at a qualitative (nominal, discrete, ordinal) or a quantitative scale.

Three aspects in particular will be discussed in this chapter, viz. (i) a causal structural analysis of systems models with binary and qualitative information with respect to the stimulus-respons relationships, (ii) an integrated statistical approach for information measured at either a qualitative or a quantitative scale, and (iii) a multivariate analysis to extract cardinal information from qualitative data.

In various environmental modelling efforts we are facing situations in which the qualitative structure of systems models have to be analysed, because no sampled data are available or because the information concerning variables is of poor quality. Some recent developments in the field of structure analysis with binary or qualitative information are therefore presented in Section 5.2, which make use of graph theory and qualitative calculus with directed graphs and signed directed graphs. These approaches deal with the analysis of the impacts between variables when they are represented by binary or qualitative levels of measurement.

Another topic mentioned in Chapter 2 focussed attention upon the coordination problems of integrating economic and environmental phenomena, and dealt with different ways of data gathering and the different levels of measurement of variables in natural sciences and social sciences (see, among others, a discussion of the different data sources in the natural sciences and the social sciences in Section 2.15). Means of information gathering may vary from controlled experiments in natural science to survey sampling in the social sciences. The corresponding levels of measurement vary from a cardinal (or metric) scale in natural sciences to a nominal and ordinal (or non-metric) scale in social sciences. An inspection of the various levels of measurement is also relevant in the framework of the operationalization of an IEM because it may be based on the integration of natural and social processes. Four kinds of measurement scales may be distinguished (in accordance with Adelman and Morris, 1974; Lammerts van Bueren, 1982; Nijkamp, 1984b), viz.:

(1) the nominal scale with a set of objects, attributes or properties from deserved variables which have been classified into distinct groups, or into distinct size classes, without restrictions on the numerical representations;

(2) the ordinal scale which meets the requirement for the nominal scale with an additional requirement that a logical ordering of magnitude exists of the events or effects measured. The variables deal with a hierarchical nature, and can be ranked from 'low' to 'high' with numbers (1,2,3,...). The difference between two ordinal figures has no numerical interpretation;

(3) the interval scale which meets the requirements for the ordinal scale and also allows determination of relative Euclidean distances between figures. The figures do not have an absolute meaning;

(4) the cardinal scale (or ratio scale) with an absolute numerical interpretation, which can be represented in a metric system.

A qualitative scale can be either nominal or ordinal, and a quantitative scale can be either interval or ratio. The four levels of measurement mentioned here form a hierarchical system, because higher order variables include all the properties of the variables measured at a lower level.

In this context, an integrated family of statistical models, called generalized linear models, is presented in Section 5.3 for linear models with variables measured at either a qualitative (or non-metric) or a quantitative (or metric) scale.

Finally, the information in systems with economic and environmental phenomena, which are analysed simultaneously will also be multivariate in nature.

Scaling methods for different levels of measurement (nominal or ordinal) are often used to identify a certain structure from an IEM system. Phenomena are multivariate in nature because various components (e.g., regions) can be characterized by various items (demographic development, infrastructural facilities, environmental structure etc.). Multidimensional procedures are then employed in order to assign numbers to each item. Scaling procedures reflect relationships among the various items, and two scaling procedures are presented in Section 5.4, viz. multidimensional scaling (MDS) (also called ordinal geometric scaling) for ordinal information, and homogeneity scaling by alternating least squares (HOMALS) for nominal information. The techniques aim at extracting cardinal information from qualitative data and have originally been developed mainly for use in psychometrics.

The mathematical and statistical approaches discussed in this chapter will be used in the chapters 6 and 7 in case of an IEM with a satellite concept of integration. This chapter will therefore be concluded with a discussion of the relevance of the statistical and econometric tools from Sections 5.2 to 5.4 and the spatial aspects discussed in Section 4.5, in the framework of the satellite concept of an IEM. This will be presented in Section 5.5.

5.2. SYSTEMS WITH BINARY, QUALITATIVE OR QUANTITATIVE INFORMATION

5.2.1. Introduction

The discussion of various IEMs in Chapter 2 showed that the data source for such models may be different in nature or even being unreliable, because of the different sources of information (surveys, national and regional censuses, expert judgments, etc.), or because of imprecise knowledge concerning impacts, or because of absence of quantified information. Table 3.1 in Chapter 3 concerning an evaluation of IEMs mentioned that the availability of information causes problems in various modules, which vary from the poor quality of information to the absence of information.

In this section some methodological tools will be discussed with respect to integrated models with a complex structure for the interpretation of the causal structure, when only limited information is available concerning the impacts between variables. Complexity is interpreted here in terms of the number of relationships within a model, i.e., it is a property of a model. The limited information of the parameters considered here is either binary (or zero/one) or qualitative (positive/negative impacts) in nature; the properties derived only depend on the causal model structure with its stimulus-response relationships, and on the qualitative information concerning impact parameters.

In terms of the interpretation and analysis of the causal structure of a model, the following measurement levels may be distinguished:

(1) A binary relationship, which indicates whether or not a certain variable has an impact on another variable. The binary relationship is denoted by zero/one information. A zero impact denotes then absence of an a priori relationship between two variables. The binary relationships denote the causal model structure. Graph-theoretical approaches are a useful tool to analyse the causal structure of models. A graph serves to represent a system of stimulus-response dependencies or a system of causal relationships. The binary relationship is based on a priori information or an a priori hypothesis concerning the causal model structure. This graph-theoretical approach concerning a binary relationship will be discussed in subsection 5.2.2;

(2) A qualitative relationship is a binary relationship with additional information on the signs of the impacts. In other words, only qualitative information about the directional relationships between variables in a model - represented by a positive (+), a negative (-) or a zero (0) impact - is available without quantitative information about the numerical values of the parameters. The so-called sign-solvability analysis of qualitative relationships, to solve a set of linear equations in a qualitative way, will be discussed in subsection 5.2.3. This sign-solvability approach belongs to qualitative calculus;

(3) A quantitative relationship with respect to the stimulus-response relationships. Path analysis may be employed when the causal relationships between variables are quantified by path coefficients, and is based on an ordinary least squares estimation procedure to estimate model parameters. The direct and indirect relationships between variables in a causal model with path analysis are discussed in subsection 5.2.4.

There are three main reasons for the treatment of binary and qualitative information concerning the impacts between variables in models, viz.:

- "Ordinarily, the economist is not in the position of having exact quantitative knowledge of the partial derivatives of his equilibrium conditions" (Samuelson, 1947, p. 26), because of the limited amount of suitable quantitative data;

- The empirical a priori information, depicted in qualitative terms, on the coefficients in a simultaneous equation system only allows to predict the effects in qualitative terms from the relevant structural system parameters (Lancaster, 1962);

- Difficulties which may arise in empirical practice to obtain precise' or exactly quantified information because of measurement problems to get the high-level information, lack of time or simply lack of money to collect the relevant data of system parameters (Nijkamp et al., 1985).

In the next three subsections the problems inherent in binary, qualitative and quantitative information will successively be dealt with.

5.2.2. Interpretation of systems with binary information

The interdependencies and hierarchical nature of the causal structure of systems will be discussed in this section by means of graph-theoretic approaches. The development of this approach is necessary because of the increase in size and mutual dependency of models including (regional) economic, demographic, and environmental aspects. The properties which may be derived from a graph do not depend on a particular normalization rule because the graph only provides information on whether or not impacts between variables are included.

Forrester (1971), for example, developed a complex world-dynamic simulation model with a large number of interconnected variables. Five major state modules were defined in the model, viz. non-agricultural investment, population, natural resources, pollution and agricultural investment. The large number of interconnected variables demonstrates the relevance of a systematic analysis of the impacts between variables.

Consider a model which consists of I explicit relations specified as

$$y_i = g_i(y,z), \quad i = 1, \ldots, I. \tag{5.1}$$

where y is a vector of endogenous variables of order $I \times 1$, and z is a vector of exogenous variables of order $J \times 1$. The structural relationships g_i ($i = 1, \ldots, I$) characterize the linkages between the endogenous variables y and the exogenous variables z, and specify the endogenous variables y_i. The causal model representation is identified by the structural relationships, and the set of endogenous variables y and exogenous variables z. The relations g_i can be linear or non-linear in nature. The information about the model structure can be expressed by a directed graph (or shortly, digraph) G. A graph is in general terms identified by a set of vertices $V = (V_1, \ldots, V_M)$ and a set of edges E, with $E = (E_1, \ldots, E_N)$. Any directed graph G is defined by the elements from V and E and $G = (V,E)$ with the set of edges which are now directed in nature. The set of vertices V - defining a graphical representation of equation (5.1) - consists of the union of the endogenous variables y and the exogenous variables z. The total number of vertices M therefore equals the sum of the number of endogenous variables I, and the number of exogenous variables J (i.e., $M = I+J$). The set E consists of the edges which represent the links between the variables y and z. The digraph G is a finite, non-empty, irreflexive and binary relationship corresponding to the set of vertices and the set of edges (see also Roberts, 1978). Such a graph can be visualized in a figure (see e.g. Figure 5.1). The information about the structure of parameters in terms of graphs can also be expressed analogously in matrix terms (Tinkler, 1977), when denoted by means of an adjacency matrix A with elements a_{ij} $(i,j =1,..,M)$. The elements from the matrix A are defined by:

$$a_{ij} = \begin{cases} 1, \text{if an edge exists from vertex } j \text{ into vertex } i \ (i,j = 1, \ldots, M) \\ 0, \text{if otherwise.} \end{cases}$$

A non-zero derivative $\partial g_i/\partial y_j$ from (5.1) indicates the existence of a causal link from variable y_j to the endogenous variable y_i. The corresponding cell-element in the adjacent matrix is equal to one. A non-zero derivative $\partial g_i/\partial z_k$ from (5.1) indicates the existence of a causal link proceeding from variable z_k to the endogenous variable y_i, which again corresponds to a cell-element in the matrix A equal to one.

A path of length n between each pair of variables is denoted in an analytical way by A^n $(n > 1)$. Thus, when some cell-element (i,j) of A^n is unequal to zero, a vertex v_i can be reached from v_j by at least one path in graph G in n steps. The cell-elements (i,j) from A^n denote the total number of paths which exist from vertices v_j into v_i.

A group of vertices within a directed graph is also called a strong component when direct or indirect links exist between all pairs of vertices which belong to the same group, and it is therefore called a maximally connected subgraph. The strong components are the basic elements for analysing model interdependencies because they denote the separate blocks with variables which are mutually dependent within a larger model. A set of vertices which is called a strong component is strongly connected; a strong component is also called a block (i.e., a maximum set of interdependent variables).

A reduced graph can then be defined in order to denote the causal links between blocks. The recursive properties can be derived from the reduced graph in order to establish a hierarchy among the blocks and to determine whether separate submodels exist. A reduced graph is defined in such a way that it does not contain any cycle (i.e., a path going from v_i to v_i in a number of steps) and consists of a unique set of arcs which define all paths (all possible direct and indirect links between points x_i and x_j). See also Ganin and Solomatin (1984); Garbely and Gilli (1984); Gilli (1984) and Gilli and Rossier (1981) for a technical discussion of the interpretation of complex systems by means of the hierarchical order form.

In order to illustrate these basic concepts, consider a three equation Keynesian economic model with the following explicit relationships:

$$C = g_1(Y,i,T)$$
$$I = g_2(Y,i)$$
$$Y = C + I,$$

with endogenous variables consumption (C), investments (I) and national income (Y), and exogenous variables interest rate (i) and taxes (T). The causal structure can be denoted in graph terms like in Figure 5.1(a) (Gilli, 1984). The graph G which is defined by the pair (V,E), consists of a set of vertices V = (C,Y,I,i,T) and a set of edges E with digraphs ((i,C), (i,I), (C,Y), (Y,C), (Y,I), (I,Y), (T,C)). The graph G has one strong component with the elements C, I and Y. The reduced graph G_r, which is derived from G by means of exclusion of all cycles, is depicted in Figure 5.1(b). The adjacency ma-

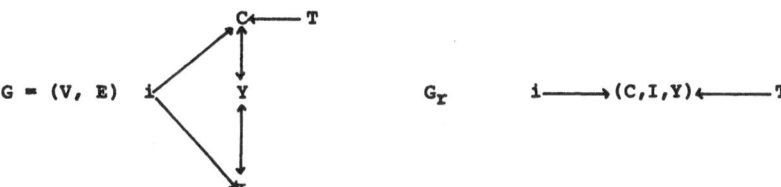

Figure 5.1(a) Graph of the three   Figure 5.1(b) Reduced graph of the
       equation model.   three equation model.

trix A(G) is an analytical equivalent to the graph representation and becomes:

$$A(G) = \begin{array}{c} \\ i \\ T \\ C \\ Y \\ I \end{array} \begin{array}{c} \begin{array}{ccccc} i & T & C & Y & I \end{array} \\ \begin{bmatrix} 0 & 0 & 0 & 0 & 0 \\ 0 & 0 & 0 & 0 & 0 \\ 1 & 1 & 0 & 1 & 0 \\ 0 & 0 & 1 & 0 & 1 \\ 1 & 0 & 0 & 1 & 0 \end{bmatrix} \end{array} \qquad A(G_r) = \begin{array}{c} \\ i \\ T \\ (C,I,Y) \end{array} \begin{array}{c} \begin{array}{ccc} i & T & (C,I,Y) \end{array} \\ \begin{bmatrix} 0 & 0 & 0 \\ 0 & 0 & 0 \\ 1 & 1 & 0 \end{bmatrix} \end{array}$$

The links between variables, or the causal relationships in a graph can be subdivided into three groups, viz. a direct causal link (or arc) between points i and j, an indirect causal link (or path) from point i to point j and an interdependent causal link (or cycle) from point i back to point i. The set of all interdependent variables which are related to some vertex i is called a strong component. The strong component of the graph G=(V,E) in Figure 5.1 (a) is the set (C,I,Y). The reduced graph in Figure 5.1(b) consists of the causal links between blocks of variables, and is derived from the graph by combination with the strong component.

The recursive causal model structure is represented by the reduced graph in

Figure 5.1(b). The corresponding adjacency matrix $A(G_r)$ is depicted above. The reduced graph shows that the endogenous variables are mutually dependent in this example, because they form a strong component.

The main advantage of the graph representation of a model is the determination of model dependencies, i.e., the interdependent blocks of the model and the hierarchical ordering of the variables in a linear model. Such a hierarchical ordering will be used in the next section for the recursive procedure to solve a model with qualitative information.

This section indicated that metric information on the impacts between variables is not always necessary to draw reliable conclusions concerning certain variables or policy aspects. Graph theory may provide a practical tool for determining, for example, the degree of interwovenness of systems, their hierarchical ordering and causality, as well as the key driving forces.

5.2.3. Interpretation of systems with qualitative information

In recent years the analysis of qualitative relations has become an increasingly important tool in impact analysis of policy modelling because of a renewed interest in qualitative calculus. The analysis of qualitative relations in economic models has been originated by Samuelson already in 1947 in order to examine - in a comparative static context - the effect on an equilibrium situation when the levels in one or more of the exogenous variables are changed. Consider the functional relationships of the form:

$$f_i (y_1, \ldots, y_n; z_1, \ldots, z_m) = 0 \quad , i = 1, \ldots, n \qquad (5.2)$$

with respect to n endogenous variables y_i ($i = 1, \ldots, n$) and m exogenous variables z_k ($k = 1, \ldots, m$).

The change in the equilibrium values of the variables y_i are determined by changes in one or more of the variables y_j and z_k and it can be obtained by differentiating the equilibrium system totally which gives:

$$\sum_{j=1}^{n} \frac{\partial f_i}{\partial y_j} dy_j + \sum_{k=1}^{m} \frac{\partial f_i}{\partial z_k} dz_k = 0. \qquad (5.3)$$

Samuelson posed the problem of equilibrium analysis in a qualitative manner because his analysis was only based on knowledge of the signs of the partial derivatives from a system like in (5.3). This set of equations can be rewritten in matrix notation as

$$Ay + b = 0, \qquad (5.4)$$

with A a matrix of order n x n with elements a_{ij} ($i, j = 1, \ldots, n$) which are equal to $\partial f_i / \partial y_j$; b a vector of order n x 1 with elements b_i equal to

$$\sum_k \frac{\partial f_i}{\partial z_k} \, dz_k.$$

Given the assumption that matrix A is non-singular, the solution of equation (5.4) is given by:

$$y = - A^{-1}b. \tag{5.5}$$

The analysis mentioned above - to obtain solution (5.5) from (5.4) - is called <u>qualitative calculus</u> if the relationships between variables are only represented in a qualitative way, i.e. when information about the sign of the impact on some response (or endogenous) variable is obtained from prior knowledge concerning the signs of the structural parameters in a model. In such cases, only qualitative information about the directional relationships between variables in a model - represented by a positive (+), negative (-) or zero (0) impact - is used without quantitative information about the numerical values of the parameters. A zero impact denotes then absence of an a priori theoretical relationship between a pair of variables, and is then analogous to a zero value of the binary relationship discussed in Section 5.2.2.

The rationale to apply a qualitative approach is that in many situations - particularly in complex multidisciplinary systems - the data base for (policy) impact analysis is unsatisfactory in order to estimate the impact parameters in an adequate way (for instance, due to lack of appropriate time-series). Especially in the case of uncertain or dynamic behaviour of a complex system, various fluctuations may occur which reflect occasional asymmetric behaviour during different time phases (for instance upswing and downswing phases in situations with structural changes), so that precise numerical values of impact parameters are unknown. An appropriate way of dealing with a weak database is qualitative calculus. Qualitative methodological tools may be used to 'quantify' the impact of policy instruments in terms of positive, negative or zero 'values'.

The step from equation (5.4) to (5.5) is called <u>sign-solvability analysis</u>, and that approach is a main stream in the field of qualitative calculus. The solution (5.5) from the set of linear equations in (5.4) is called sign-solvable when the cell-entries of the solution $y = -A^{-1}b$ (with vector-elements either +, - or 0) are defined in a unique way, provided both matrix A and vector b contain qualitative information concerning impact levels between variables.

Full sign-solvability of equation (5.4) denoted in terms of (5.5), holds, if and only if both the inverse of matrix A as well as the inverse of A multi-

plied with vector b are determined uniquely. Fortunately there is a number of
matrix operations which are relevant in empirical situations for the analysis
of the conditions of sign-solvability because they do not affect the analysis
of sign-solvability (see also Lancaster, 1962), viz.:

(i)   permutation of any two rows of both matrix A and vector b. This opera-
      tion only changes the order in which the equations are written;

(ii) permutation of any two columns from matrix A. This operation only chan-
      ges the order of the variables;

(iii) reversal of all signs in any row of both A and b. This operation multi-
      plies both sides of an equation with a factor -1;

(iv) reversal of all signs in any column of matrix A. This operation mul-
      tiplies a variable by -1.

These matrix operations may simplify the analysis of sign-solvability, which
will be shown below when the conditions of sign-solvability are discussed.
The row and column manipulations (i), (ii) and (iii) can be carried out without affecting the solution in (5.5), and the final operation implies the
sign reversal from that particular variable.

Necessary and sufficient conditions for full sign-solvability of equation
(5.4), with matrix A considered to be non-singular, have been formulated by
Bassett et al., (1968). The conditions make use of graph-theoretic methods,
with impacts between variables denoted by signed digraphs (directed graphs
with either a positive or a negative sign), and are:

(a) the diagonal elements of the matrix A are all negative, i.e. $a_{ii} < 0$
    for all $i = 1, \ldots, n$;

(b) all cycles in the matrix A with length at least two need to be non-posi-
    tive;

(c) $b \leqslant 0$: the elements of vector b need to be non-positive;

(d) if some $b_k < 0$, then every path from vertex k to vertex i is non-posi-
    tive for $i \neq k$.

The row and column operations (i) to (iv) are useful for the investigation of
the conditions of sign-solvability, because the first condition of sign-solvability can be met after one or more row or column permutations and sign re-
versals.

Conditions (a) and (b) of sign-solvability are necessary and sufficient to
determine the inverse of the matrix A in a unique way, while the inclusion of
the other conditions guarantees a unique sign of the cell-entries from $A^{-1}b$.

The conditions of sign-solvability are now discussed using the following
illustration. Consider the analytical representation from a set of four linear equations with qualitative information about the impacts between varia-

bles:

$$
\begin{bmatrix} - & + & - & 0 \\ - & - & 0 & 0 \\ 0 & 0 & - & + \\ 0 & 0 & - & - \end{bmatrix}
\begin{bmatrix} y_1 \\ y_2 \\ y_3 \\ y_4 \end{bmatrix}
+
\begin{bmatrix} 0 \\ - \\ 0 \\ 0 \end{bmatrix}
=
\begin{bmatrix} 0 \\ 0 \\ 0 \\ 0 \end{bmatrix}
$$

The impacts between variables from this system are represented in an equivalent way in Figure 5.2 by means of signed digraphs of (A,b).

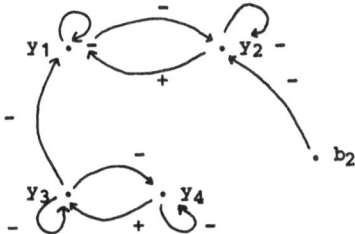

Figure 5.2. A graph representation of a qualitative model.

All main diagonal elements from matrix A are negative, and the first condition of sign-solvability holds. The length of a cycle represents the number of terms which appear in the cycle. Thus, the graph in Figure 5.2 has two cycles with length at least two, viz. y_1-y_2-y_1 and y_3-y_4-y_3. Both cycles with length two are negative, so that the second condition of sign-solvability also holds. The elements of vector b are non-positive, because vertex b_2 has an outgoing edge which is negative in sign. The path b_2-y_2-y_1 is negative in sign and the final condition of sign-solvability therefore also holds in Figure 5.2. Thus, all conditions for sign-solvability hold in this example and the solution in equation (5.5) now becomes:

$$
\begin{bmatrix} y_1 \\ y_2 \\ y_3 \\ y_4 \end{bmatrix}
=
\begin{bmatrix} - & + & - & 0 \\ - & - & 0 & 0 \\ 0 & 0 & - & + \\ 0 & 0 & - & - \end{bmatrix}^{-1}
\begin{bmatrix} 0 \\ + \\ 0 \\ 0 \end{bmatrix}
=
\begin{bmatrix} - & - & + & + \\ + & - & - & - \\ 0 & 0 & - & - \\ 0 & 0 & + & - \end{bmatrix}
\begin{bmatrix} 0 \\ + \\ 0 \\ 0 \end{bmatrix}
=
\begin{bmatrix} - \\ - \\ 0 \\ 0 \end{bmatrix}
$$

The changes in the variables y_1 and y_2 thus become negative, while the other two variables do not change from their equilibrium position, given the model structure represented by matrix A and the signs of the exogenous variable in vector b.

The sign-solvability approach can be interpreted as a kind of overall __sensitivity analysis__ in the following way. If it is possible that equation (5.4) can be solved for y in a unique way with a vector of signs as the solution, the solution will, up to their signs, hold for all possible cardinal values of the matrix A and vector b.

The main developments in the field of sign-solvability analysis took place in mathematics (see also Greenberg and Maybee, 1981, which presented the results of a symposium on Computer Assisted Analysis and Model Simplification held in 1980, and a review paper on recent developments in the field of sign-solvability analysis by Maybee and Voogd, 1984), economics (see among others, Lady, 1983; Ritschard, 1983) and ecology (see also Brouwer and Nijkamp, 1985, as well as Jeffries, 1974, for a discussion of asymptotic stability of a set of linear differential equations with qualitative information by making use of signed directed graphs). Asymptotic stability with qualitative information on a set of linear differential equations $\frac{dy}{dt} = Ay$ can also be interpreted in terms of graph theory, viz. in a way analogous to sign-solvability in (5.4). Asymptotic stability of linear differential equations with qualitative information concerning the impacts between species also belongs to the analysis of qualitative relations.

The conditions for sign-solvability discussed above originate from mathematics. They have - in the framework of economic modelling - been further examined inter alia for the well known national model of the USA developed by Klein (see also Brouwer and Nijkamp, 1986c). The latter exercise demonstrated that sign-solvability did not even hold for such a small dynamic national economic model with six equations. Consequently it may be expected that other usually larger models with qualitative information will not provide more satisfactory results.

A disadvantage of sign-solvability analysis with pure qualitative information is that the four conditions for sign-solvability are very severe. However, the inclusion of additional tools to sign-solvability analysis may give a more useful methodology for a qualitative approach in empirical applications. Some adjusted tools of qualitative approaches and recently developed research directions will therefore be discussed in this section, with a particular view to the use of the sign-solvability approach in policy modelling based inter alia on integrated environmental models (see also Brouwer and Nijkamp, 1986e), viz.:

(i)   the use of matrix decomposition and matrix permutation procedures;

(ii) the use of plausible parameter restrictions which may be inferred on theoretical grounds, or on a priori knowledge;

(iii) the use of a stepwise procedure to include parameter values from one or more equations which are based on a priori information. This stepwise procedure makes a distinction between a so-called top-down and a bottom-up approach;

(iv) the use of recently developed computer algorithms for the analysis of qualitative linear systems.

First, a number of __matrix permutation procedures__ and __matrix decomposition procedures__ may be used for the analysis of sign-solvability. Such matrix procedures imply a resettlement of variables and equations, and it will give a hierarchical model representation. Reducibility of matrices is an essential characteristic in the case of matrix permutation procedures (Maybee, 1981). A matrix A is called reducible when a permutation matrix P exists, in order to reverse rows and columns of the matrix A, so that A will be transformed into A* with:

$$A^* = \begin{bmatrix} A_{11} & 0 \\ A_{21} & A_{22} \end{bmatrix},$$

where both matrices A_{11} and A_{22} are square matrices, and 0 is a zero-matrix. The transformation of A into A* is obtained by:

$$A^* = P A P^T, \tag{5.6}$$

with P a permutation matrix to reverse rows and columns from matrix A into matrix A*. When matrix A is reducible, the matrix can be decomposed in the sub-matrices from matrix A*, while next the sign-solvability approach can be dealt with in two steps, because (5.4) can be rewritten as a recursive system:

$$\begin{bmatrix} A_{11} & 0 \\ A_{21} & A_{22} \end{bmatrix} \begin{bmatrix} y_1 \\ y_2 \end{bmatrix} + \begin{bmatrix} b_1 \\ b_2 \end{bmatrix} = \begin{bmatrix} 0 \\ 0 \end{bmatrix} \tag{5.7}$$

or:

$$\begin{aligned} A_{11} y_1 &= -b_1 \\ A_{21} y_1 + A_{22} y_2 &= -b_2. \end{aligned} \tag{5.8}$$

The necessary and sufficient conditions of sign-solvability of the vector y_1 can be analysed independently from vector y_2. The linear model in (5.7) is a hierarchical form of the equivalent model in (5.4), and in such hierarchical cases sign-solvability can be investigated for only the first part of model (5.7). Vector y_1 may thus be sign-solvable irrespective of whether vector y_2 is sign-solvable. The variables are ordered hierarchically in (5.7), and the __model__ can then be __solved recursively__ by starting with the elements from vector y_1.

The second extension of sign-solvability concerns the use of plausible information on parameter values. Such plausible information is based on logical,
empirical or theoretical evidence and may also be an important research strategy in qualitative analysis. An originally not sign-solvable system may become sign-solvable when plausible information is introduced. An example of
such plausible information in national-economic models is the share of consumption in national income which is not only positive in sign but should
also have plausible values in the range between zero and one. Such plausible
quasi-quantitative information may lead to sign-solvable systems which are
not sign-solvable when only purely qualitative information was used. One of
the major problems in practical applications with purely qualitative information is caused by the severe restrictions inherent in identifying unambiguous
solutions for sign-solvability. In this regard, additional a priori information may lead to at least partial sign-solvable systems which are otherwise
not solvable in a purely qualitative sense (see also Brouwer and Nijkamp,
1986e for a discussion of sign-solvability analysis applied to a dynamic
simulation model of urban decline which makes use of a mixture of qualitative
and quantitative information concerning impact parameters).

The third extension deals with the use of a mixture of qualitative and quantitative information which may be distinguished into a top-down and a bottom-
up approach. Both approaches are stepwise procedures to insure that a qualitative system will become sign-solvable in a number of steps by a sequential
introduction of numerical information on parameter values. The top- down
approach (also called forward selection) implies that all equations are assumed to be represented in qualitative terms in the first step, so that we
then may identify which and how many equations have to be estimated in a
quantitative way (i.e., how many plausible parameter restrictions are necessarily to be included), in order to make a set of equations sign-solvable.

The bottom-up approach (also called backward elimination) starts with a com-
pletely estimated model and attempts then to identify which and how many
equations may be specified in qualitative terms in order to still guarantee
sign-solvability. A main advantage of both approaches is that complex integrated models may become sign-solvable when a mixture of qualitative and
quantitative information is used and the precision of the quantitative information on some parameters is doubted. The top-down/bottom-up approach may
also make use of plausible information on parameter values.

Finally, a FORTRAN computer program has been developed at the University of
Geneva by Ritschard and Gilli to solve qualitative linear systems by means of
a block recursive decomposition procedure (see also Gilli, 1984 and Ritschard, 1980). The hierarchical representation of a linear set of equations

in (5.4), which has then to be solved recursively equation by equation, may however become problematic for computational reasons, especially in case of complex dynamic simulation models.

The conclusion can be drawn from the above that a qualitative approach to integrated environmental modelling is a useful mathematical tool to understand the global functioning and coherence of a model (Royer and Ritschard, 1984). A qualitative approach concerning the impacts between variables is a useful approach in integrated environmental analyses when the data are of poor quality or when no sampled data are available at all. The sign-solvability approach determines the change in qualitative terms in the response variable due to changes in the stimulus variables. The sign-solvability approach may be extremely relevant, as this method may be able to predict the qualitative (sign) impact of an exogenous change, even if reliable cardinal information about impact coefficients is not available. This qualitative approach may be an appropriate complementary tool to conventional econometric estimation techniques and simulation procedures.

The use of sign-solvability analysis for urban systems models was presented by Brouwer and Nijkamp, 1986e, by means of a dynamic simulation model of urban decline for The Hague in the Netherlands. Parameter estimation and model validation was problematic because of the a- symmetric pattern of urban evolution and the lack of appropriate, reliable quantitative information. Given the insufficiently reliable data base for estimating the model in a conventional way, qualitative calculus was used for a linear difference equations model in order to infer conclusions regarding the direction of impacts of policy variables.

5.2.4. Interpretation of causal systems with quantitative information

The model structure of systems was analysed in subsections 5.2.2 and 5.2.3 by making use of only binary or qualitative information and of graph-theoretic tools. The model structure was represented by directed graphs and signed directed graphs successively. Causal systems, which are essentially based on multiple regression analysis when variables are measured on an interval or ratio scale and with an additional assumption on causal relationships between the independent and dependent variables, will be interpreted now for quantitative information by means of path analysis. Causal processes are reflected by an arrow diagram and may facilitate a more clear statement of hypotheses concerning the presence or absence of the linkages between variables. A causal model is defined as a system of considerations about cause-and-effect-relationships. The analysis of binary relations (in Section 5.2.2) and the

analysis of qualitative relations (in Section 5.2.3) both explicitly represent the type and direction of the causal relationship between variables.
However, the causal model structure itself will be discussed in this section
by means of path analysis. Path analysis represents an extended approach of
causal modelling, as it estimates the magnitude of the linkages between variables, rather than focusing upon their presence or absence. An overview of
structural equations models with applications in the field of human geography
is presented by Cadwallader (1986). The standardized partial regression coefficients can be interpreted as direct effects. A standardized regression
coefficient is obtained when each variable is considered to have a zero mean
and standard deviation of one (see also Duncan, 1971 for a discussion of path
analysis as linear causal models). If a cause-and-effect or a stimulus-response relationship is considered, then a significant parameter is interpreted
as the main effect. The basic elements of path analysis are now discussed by
means of an example presented in Figure 5.3.

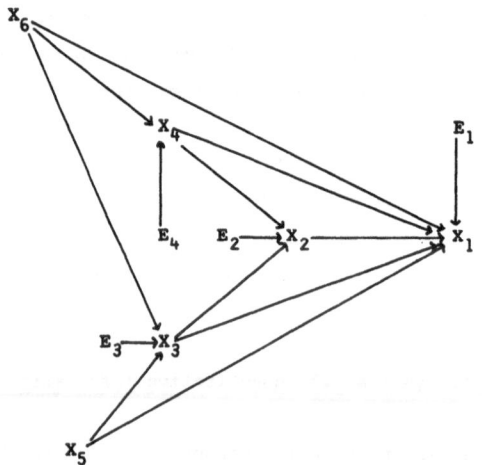

Figure 5.3. A path-diagram.

The variable at the head of one or more arrows is considered to be a function
of those variables at the tails of the same arrows. Variable X_1 in Figure 5.3
for example, is interpreted as being a linear additive function of the variables X_2 to X_6. The variables X_5 and X_6 in Figure 5.3 are considered to be
exogenous in nature. The path coefficients p_{ij} denote the path from
variable X_j to variable X_i.
The model which corresponds to the path-digram in Figure 5.3 becomes:

$$X_1 = p_{12} X_2 + p_{13} X_3 + p_{14} X_4 + p_{15} X_5 + p_{16} X_6 + p_1 E_1$$
$$X_2 = \qquad\quad p_{23} X_3 + p_{24} X_4 \qquad\qquad\qquad\qquad + p_2 E_2 \qquad\qquad (5.9)$$
$$X_3 = \qquad\qquad\qquad\qquad p_{35} X_5 + p_{36} X_6 + p_3 E_3$$
$$X_4 = \qquad\qquad\qquad\qquad\qquad\qquad p_{46} X_6 + p_4 E_4$$

The path coefficients p_{ij} are estimated with a least squares estimation procedure, and are the standardized regression coefficients. The path coefficients in the first equation from (5.9) for example, are obtained by regression of X_1, from X_2 to X_6. If R_1^2 is defined to be proportion of variance from X_1, then the square root of $1-R_1^2$ gives the value from the 'error' variable. The following assumptions are made to use the ordinary least squares regression procedure (see also Macdonald, 1977):

(1) a correct specification of the causal relationships between variables which also includes the assumption of linear additive relationships. The condition of linear additivity means that a unit change of variable X_2 in equation (5.9) has the same effect on variable X_1, irrespective of the value of X_2;

(2) independence of the error terms E_i from the causal variables X_j. The error terms are also uncorrelated because of the assumed linear additivity;

(3) homoscedasticity (or constant variance) of the error terms;

(4) the variables are measured at the interval or ratio level of measurement.

The influence between two variables can be subdivided into **direct effects** and **indirect effects**. Consider for example the effect between the variables X_1 and X_5. The direct effect is equal to p_{15}. However, indirect effects also exist between these two variables, viz. with variable X_3 as intermediate variable (a path exists $X_5 \rightarrow X_3 \rightarrow X_1$ and a path exists $X_5 \rightarrow X_3 \rightarrow X_2 \rightarrow X_1$). The sum total of indirect effects is equal to $p_{35}p_{13} + p_{35}p_{23}p_{12}$.

Applications of path analysis with variables measured on a nominal scale and information obtained from survey samples are presented by Asher (1983) concerning the transmission of party identification across generations; by Hall and Taylor (1983) concerning the attitudes toward mental health facilities; by Leitner and Wohlschlägl (1980) concerning satisfaction from recreationers; by Leitner et al. (1985) concerning the factors which influence spatial segregation, structural integration and behavioral assimilation of migrant workers in Vienna, and by Takeuchi et al., (1982) concerning the intervening role of mass media exposure for predicting family planning knowledge.

The **statistical significance of any path coefficient** in an equation may be tested by means of a two-sided test-statistic which follows a student-distribution (the parameter estimate related to the corresponding standard error has asymptotically a student-distribution, with degrees of freedom equal to n-p-1, with n the number of observations and p the number of parameters in an equation). The _statistical significance of two or more path coefficients_ can be tested with a test-statistic which has a Fisher-distribution. If we define RSS_u and RSS_r as the sum of squared residuals of the unrestricted model and the restricted model respectively, n as the number of observations, p as the number of parameters in the unrestricted model and k the number of para-meter restrictions (for example, the number of parameters considered to be equal to zero), the test-statistic becomes:

$$\frac{(RSS_r - RSS_u) / k}{RSS_u / (n-p-1)} , \qquad (5.10)$$

with an F-distribution with (k, n-p-1) degrees of freedom.

5.3. STATISTICAL MODELS WITH QUALITATIVE AND QUANTITATIVE DATA

5.3.1. Introduction

Traditionally the statistical modelling approaches in geographic, regional-economic, environmental and transportation analysis, for example, show metric or quantitative data measured at the interval or ratio level. Examples of frequently used approaches are regional and interregional input-output analy-sis, cost-benefit analysis, multi-objective decision modelling, spatial equi-librium analysis, etc. This research tradition with a major emphasis on high-quality information followed the standard practice in natural science where information is collected under well defined conditions from controlled exper-iments. Interval and ratio scales and roughly normal distributions seemed to be the rule rather than the exception. Unfortunately, this tradition too often ignores the realities of regional-economic, social and ecological re-search where the information available for statistical modelling purposes will frequently have been obtained from other sources, such as expert judge-ments, survey-sampling or panel-studies, and will frequently be non-metric (qualitative, discrete or categorical) in nature. 'Measurement of properties of human behaviour poses many problems such as instability, bias by reactivi-ty, inherent discreteness and sometimes nominal or ordinal measurement level' (Molenaar, 1985, p. 171). Adequate methodological tools for dealing with this type of information have been developed in the past two decades. Bartholomew for example mentioned that 'for too long social scientists have had to manage with methods designed for use in the natural sciences in which the variables are well defined and readily capable of measurement. In the last decade or two the balance has been somewhat restored by the development of such things as the log-linear model' (Bartholomew, 1980, p. 127).

The aim of this section on statistical models with qualitative and quantitative data is to describe the methodological advances which have taken place over the past twenty years in this research area, and to discuss their relevance in regional-economic and environmental research. In subsection 5.3.2 the qualitative statistical models are specified within a family of generalised linear models (GLMs), a statistical framework developed by Nelder and Wedderburn. Particular attention will be placed in subsection 5.3.3 upon log-linear models and logistic/logit models for exploratory and explanatory data analysis of multidimensional contingency tables. Model estimation, model assessment and model selection procedures are discussed in subsection 5.3.4. Multidimensional contingency tables with cross-classified variables may also be analysed in terms of graphs with stimulus-response relationships. This will be further elaborated in subsection 5.3.5.

5.3.2. Classification and integration of statistical approaches

Continuous variables are variables measured at the high level of an interval or ratio scale and are quantitative in nature. Categorical variables differ from continuous variables because they are measured at a low level of measurement, i.e. a nominal or an ordinal scale. A distinction can be made between three types of categorical variables, viz.:

(i)   dichotomous variables (e.g., presence or absence, yes or no, male or female);

(ii)   unordered polytomous variables (e.g., region A, region B, region C);

(iii)   ordered polytomous variables (e.g., high income, middle income, low income).

Dichotomous and unordered polytomous variables are measured at a nominal scale because the variables or phenomena concerned can only be distinguished by their name or attribute. In order to classify such variables, numbers as well as other symbols can be used.

The ordered polytomous variables can be ranked from low to high (implying that they are measured at the ordinal scale) because it is known whether some classified variable has a higher or a lower value than another one. An example of an ordered polytomous variable is a regional environmental quality profile with values 'bad', 'good' and 'worse' or represented in an analytical way by (1,2,3). Differences between ordinal values do not have a numerical meaning.

One possible framework for understanding the range and interrelationships of
these new methods for analysing qualitative data is that shown in Table 5.1.
(See also Wrigley, 1979; Wrigley and Brouwer, 1986; Fischer and Nijkamp,
1985).

Table 5.1. A classification of statistical problems.

Explanatory variables

		Continuous	Mixed	Categorical	
Response variable	Continuous	(a)	(b)	(c)	
	Categorical	(d)	(e)	(f)	(g)

Cell (g) has been separated from cells (a) to (f), because it includes the
models for qualitative variables when no prior causal structure has been
considered.

Table 5.1 shows a classification of statistical problems on the basis of the
type of response (or dependent) and explanatory (or independent) variables
involved. This table is classified according to a continuous, a mixed or a
categorical level of variables, and it has two features. First, it shows that
the way of dealing with qualitative data becomes more important when moving
from cell (a) to cell (f) and (g). Second, the classical statistical models
with metric data (e.g. regression models, dummy variable regression models)
in the first row of the table are linked with recently developed methods for
qualitative data analysis (e.g. log-linear models, logistic/logit and probit
regression models) in cells (d) to (g) (see also Wrigley, 1985).

Implicit in the structure of Table 5.1 is the fact that the statistical models for qualitative and quantitative data are part of a unified family of
linear statistical models. The linkage between qualitative/categorical data
models and conventional quantitative/continuous data models can be formalized
by regarding the models as members of Nelder and Wedderburn's family of 'generalized linear models' (GLMs) (Nelder and Wedderburn, 1972).

A GLM can be expressed in this form (see also McCullagh and Nelder, 1983;
Nelder, 1985):

$$y_i = \mu_i + \varepsilon_i \quad , \quad i = 1, \ldots, N \tag{5.11}$$

with

> y_i a response variable which is assumed to originate from the
> exponential family of probability distributions;
>
> μ_i is the expected value of y_i, i.e. $\mu_i = E(y_i)$;
>
> ε_i is a randomly distributed error term.

The GLM in equation (5.11) can be characterized as a linear model for N independent univariate variables y_i and consists of the sum of a systematic
component μ_i and a random component ε_i.

The explanatory variables are denoted by X_{ik} (i=1,..., N; k=1, ..., K).
The variables X_{ik} are the measurements of the K independent variables
and they are thought to influence the variation in the response variable,
y_i, in equation (5.11). The explanatory variables can be summarized in the
structure of the so called 'linear predictor' (η_i) which takes the form:

$$\eta_i = \sum_k \beta_k X_{ik},$$
(5.12)

with β_k as the parameters to be estimated. The linear predictor (for which
a linear additivity hypothesis is considered in the GLM approach) can then be
related to the expected value of y_i by the so-called 'link-function' g

$$\eta_i = g(\mu_i),$$
(5.13)

or, in an analogous way in terms of the inverse of the link function

$$\mu_i = g^{-1}(\eta_i).$$
(5.14)

The link-function g is a monotonic twice differentiable function.

Substitution of the linear predictor and the inverse link function into equation (5.11) implies that a generalized linear model can also be written in
the form:

$$y_i = g^{-1}(\eta_i) + \varepsilon_i \qquad i = 1, ..., N$$
(5.15)

or, alternatively:

$$y_i \propto F \left(g^{-1}\left[\sum_{k=1}^{K} \beta_k X_{ik} \right] \right),$$
(5.16)

where F denotes the assumed exponential family of probability distributions.

In order to define some GLM, the link-function, linear predictor and error
distribution should be specified.

Some examples of GLMs in the field of qualitative statistical modelling (with
a discrete probability distribution) are given in Table 5.2 (see also O'Brien
and Wrigley, 1984).

Table 5.2 shows that a wide range of statistical models may be expressed in

Table 5.2. Examples of generalized linear models.

Model	Link-function	Error distribution
Linear regression	Identity: $\eta = \mu$	Normal
ANOVA	Identity: $\eta = \mu$	Normal
Logistic/logit regression	Logit: $\eta = \log_e \dfrac{\mu}{1 - \mu}$	Binomial/multinomial
Symmetric log-linear	Logarithmic: $\eta = \log_e \mu$	Poisson
Asymmetric log-linear	Logit: $\eta = \log_e \dfrac{\mu}{1 - \mu}$	Binomial/multinomial

GLM terms, e.g., the classical linear regression with a normal distribution function for continuous data linear models and also many statistical distributions for qualitative data linear models.

5.3.3. Log-linear models and logit models for contingency table analysis

A contingency table with nominal data is obtained when two or more variables are classified according to some prior chosen classification scheme giving two-way or multiway contingency tables. The modern view on contingency table analysis is influenced by a discussion between Yule and Pearson during the turn of the century (Fienberg, 1980). Pearson interpreted contingency tables in terms of categories based on continuous variables which follow a normal distribution. Yule considered categorical variables as based on a well defined category structure and he developed some measures of association between pairs of variables for a 2 x 2 contingency table, measured at a nominal scale and with discrete classes. The current points of view regarding contingency table analysis are mainly in agreement with the approach advocated by Yule.

In cell elements (f) and (g) of Table 5.1, all variables are categorical in nature. Sample data may be represented in the form of two-dimensional or multidimensional contingency tables which result from the cross-classification of categorical variables.

In the case of cell (g) from Table 5.1, no distinction has been made between explanatory and response variables. The contingency table will therefore be called a symmetric table. In the case of cell (f) from Table 5.1 a distinction has been made between explanatory and response variables, and the cor-

Table 5.3. An I x J two-dimensional contingency table.

variable A	j =	variable B 1	2	...........J	Total
	i = 1	n_{11}	n_{12}	...........n_{1J}	n_{1+}
	i = 2	n_{21}	n_{22}	...........n_{2J}	n_{2+}
	i = I	n_{I1}	n_{I2}	...........n_{IJ}	n_{I+}
	Total	n_{+1}	n_{+2}	...........n_{+J}	n_{++} = N

responding contingency table will be called an asymmetric contingency table.
Table 5.3 shows a two-dimensional I x J table with sample size N, and n_{ij}
denotes the observed cell-frequency for cell (i,j). The elements n_{i+}
(i=1, ..., I), and n_{+j} (j=1, ..., J) denote the i-th row sum and j-th
column sum respectively.

An extremely powerful exploratory method for the analysis and decomposition
of a structure from sets of data in contingency tables is log-linear modelling, in which main effects and interaction effects serve to identify patterns of association between the cross-classified variables. Patterns of
association between such variables are specified and quantified by log-linear
models. The most general log-linear model for representing data from a table
in linear additive components, which is linear in the (natural) logarithms of
the expected cell-frequencies, m_{ij}, becomes (see also Bishop et al.,
1975):

$$\log_e m_{ij} = \log_e E(n_{ij}) = \lambda + \lambda_i^A + \lambda_j^B + \lambda_{ij}^{AB} \quad ;i = 1,..,I; \; j =1,..,J. \qquad (5.17)$$

All information about the structure of the contingency table (association and
interaction) is contained in equation (5.17). The superscripts refer to the
variables involved and the subscripts to the categories of the variables.

The parameter λ is the grand- mean effect, λ_i^A and λ_j^B are the
main-effects of variables A (for the i-th category) and B (for the j-th category) and λ_{ij}^{AB} is the first-order interaction effect between variables A and B (see also Birch, 1963). A family of log-linear models may be
specified when different parameters in equation (5.17) are set equal to zero.
The equations (5.18) to (5.21) are called the <u>hierarchical</u> set of log-linear
models for a two-dimensional contingency table, because inclusion of any

higher-order parameters means that all related lower-order parameters are also included.

$$\log_e m_{ij} = \lambda + \lambda_i^A + \lambda_j^B \qquad (5.18)$$

$$\log_e m_{ij} = \lambda + \lambda_i^A \qquad (5.19)$$

$$\log_e m_{ij} = \lambda + \lambda_j^B \qquad (5.20)$$

$$\log_e m_{ij} = \lambda \qquad (5.21)$$

The main effect (or prevalence effect) expresses the contribution to the data structure, and the prevalence of the variable concerned. The interaction effect parameters express the effect of the inclusion of interactions between two or more variables.

The number of parameters in equation (5.17) is equal to $IJ + I + J + 1$ which is larger than the number of cell-elements. Parameter constraints must then be imposed to identify the parameters in order to obtain a number of IJ independent parameters, which is equal to the number of cell-elements.

Two common parameter restrictions are:

(a) the 'centred-effect' reparametrisation system

$$\sum_{i=1}^{I} \lambda_i^A = \sum_{j=1}^{J} \lambda_j^B = \sum_{i=1}^{I} \lambda_{ij}^{AB} = \sum_{j=1}^{J} \lambda_{ij}^{AB} = 0, \qquad (5.22)$$

which is equivalent to the analysis-of-variance constraints. The main-effect parameters λ_i^A and λ_j^B and the interaction effect λ_{ij}^{AB} are deviations from the overall mean-effect λ.

(b) the 'cornered-effect' reparametrisation system

$$\lambda_1^A = \lambda_1^B = \lambda_{1j}^{AB} = \lambda_{i1}^{AB} = 0, \quad i = 2, \ldots, I, \; j = 2, \ldots, J. \qquad (5.23)$$

The parameter λ represents the logarithm of the expected frequency in the 'anchor' or base cell (e.g. cell-element (1,1) in (5.23)), while λ_i^A, λ_j^B and λ_{ij}^{AB} ($i = 2, \ldots, I, \; j = 2, \ldots, J$) are interpreted as deviations from the 'anchor' cell-value or the frequency in the basic cell.

The advantage of the centred-effect reparametrisation is that the parameters can easily be interpreted, whereas the advantage of the cornered-effect reparametrisation is the independence of the observations from the categories.

To the linear model in equation (5.17) may be referred now as the <u>saturated log-linear model</u> for a two-dimensional contingency table, because it is linear in the logarithms of the expected cell-frequencies and because it satis-

$$\log_e \frac{p_{r/i}}{p_{R/i}} = \underset{\sim}{x}_i' \underset{\sim}{\beta}_r' \qquad \begin{array}{l} r = 1, \ldots, R - 1 \\ i = 1, \ldots, N \end{array} \qquad (5.31)$$

The mathematical equivalence between log-linear models for cell (f) problems from Table 5.3, and linear logit models can easily be shown for a 2xJxK contingency table with variable A as the response variable and variables B and C as explanatory variables (see also Fienberg, 1980). The saturated log- linear model in that case becomes:

$$\log_e m_{1jk} = \lambda + \lambda_j^B + \lambda_k^C + \lambda_{jk}^{BC} + \lambda_1^A + \lambda_{1j}^{AB} + \lambda_{1k}^{AC} + \lambda_{1jk}^{ABC} \qquad (5.32)$$

and

$$\log_e m_{2jk} = \lambda + \lambda_j^B + \lambda_k^C + \lambda_{jk}^{BC} + \lambda_2^A + \lambda_{2j}^{AB} + \lambda_{2k}^{AC} + \lambda_{2jk}^{ABC} \qquad (5.33)$$

The logarithm from the odds-ratio can be obtained from (5.32) and (5.33):

$$\log_e \frac{m_{1jk}}{m_{2jk}} = (\lambda_1^A - \lambda_2^A) + (\lambda_{1j}^{AB} - \lambda_{2j}^{AB}) + (\lambda_{1k}^{AC} - \lambda_{2k}^{AC}) + (\lambda_{1jk}^{ABC} - \lambda_{2jk}^{ABC}) \quad (5.34)$$

Consider a 'centred-effect' parameter constraint system like (5.22) and (5.24) which implies that (5.34) can be rewritten as

$$\log_e \frac{m_{1jk}}{m_{2jk}} = 2\lambda_1^A + 2\lambda_{1j}^{AB} + 2\lambda_{1k}^{AC} + 2\lambda_{1jk}^{ABC}$$
$$= w + w_j^B + w_k^C + w_{jk}^{BC} \qquad (5.35)$$

which is a linear logit model with categorical variables B and C as the explanatory variables (see also Grizzle et al., 1969).

The structure of log-linear models and linear logit models for qualitative data has been discussed in this section. Altogether, the conclusion can be drawn that log-linear models are a powerful tool in analysing the structure and interaction effects of qualitative information from economic, social or environmental data sources. Some model estimation, model assessment and model selection procedures for qualitative statistical models will be discussed in the next subsection.

5.3.4. Estimation, assessment and selection for qualitative statistical models

The parameter estimation procedures which are appropriate for qualitative statistical models can be subdivided into three main classes (Wrigley and Brouwer, 1986; Imrey et al., 1981):

(i)   iterative proportional fitting methods;

(ii)   weighted least squares (WLS) methods;

(iii)   function maximization techniques, e.g. Newton-Raphson, Davidon-Powell.

The _iterative proportional fitting procedure_ gives a proportional adjustment of a row and a column of a contingency table to obtain estimated expected cell-frequencies. Consider for example model type 2 from Table 5.5. Because of the nature of such hierarchical log-linear models, the minimum set of sufficient statistics in the example consists of the marginal categories n_{ij+}, n_{i+k} and n_{+jk}. The iterative scaling procedure which starts with the initial estimates of expected frequencies

$$m_{ijk}^{(0)} = 1 \quad \text{for all i, j and k,}$$

and the constraints from the minimum set of sufficient statistics give the following stage s of the iterative procedure (see also Fienberg, 1970):

$$m_{ijk}^{(3s+1)} = m_{ijk}^{(3s)} \frac{n_{ij+}}{m_{ij+}^{(3s)}} \quad \text{for all i, j, k}$$

$$m_{ijk}^{(3s+2)} = m_{ijk}^{(3s+1)} \frac{n_{i+k}}{m_{i+k}^{(3s+1)}} \quad \text{for all i, j, k} \tag{5.36}$$

$$m_{ijk}^{(3s+3)} = m_{ijk}^{(3s+2)} \frac{n_{+jk}}{m_{+jk}^{(3s+2)}} \quad \text{for all i, j, k.}$$

With each new fit in (5.36), the previous adjustments are lost. Hence, it is necessary to begin a new cycle of the iteration procedure. The iterative procedure is continued until the desired accuracy will be achieved. Convergence is obtained when only an arbitrary small difference between the estimated marginal totals and the specified marginal totals remains.

The computer programs ECTA (Everyman's Contingency Table Analysis) (see also Fay and Goodman, 1975) and BMDP Routine P4F (Biomedical Computer Programs, P series) (see also Brown, 1981) both make use of an iterative proportional fitting procedure to produce maximum likelihood estimates of the expected cell-frequencies, log-linear model parameters and standard errors. In case of logistic/logit models, both function maximization techniques and weighted least squares can, theoretically, be used. However, in case of cell (f) problems in Table 5.1., in which _all_ explanatory and response variables are categorical, the non-iterative WLS procedure is a widely used approach. The WLS parameter estimates can be obtained by minimization of the quadratic form:

$$(\bar{L} - L)' \ V_{\bar{L}}^{-1} \ (\bar{L} - L), \tag{5.37}$$

with

$$L = X \ \beta \ , \tag{5.38}$$

where $\bar{L}$ is a vector of observed logit values, and $V_{\bar{L}}^{-1}$ a matrix of weights (see also Wrigley, 1980). The WLS parameter estimates β in the equations (5.28) and (5.31) then become:

$$\underset{\sim}{\beta} = (X' \ V_{\bar{L}}^{-1} \ X)^{-1} \ X' \ V_{\bar{L}}^{-1} \ \bar{L} \ . \tag{5.39}$$

The non-iterative WLS procedure assumes that the data to be modelled have arisen from an underlying product multinominal sampling model, and depends upon the availability of moderate and large size samples for each combination of categories from the explanatory variables. In case of cells (d) and (e) problems in Table 5.1, this latter condition is rarely satisfied and the non-iterative WLS procedure is rarely used for such problems, while direct function maximization techniques (numerical optimization procedures) or the iterative WLS methods from Nelder and Wedderburn are universally adopted procedures. Both produce maximum likelihood estimates (see also Nelder and Wedderburn, 1972 for a definition of parameter estimates with an iterative WLS procedure).

Applications of linear-logit and log-linear models in the field of geography, spatial economics, psychometrics and ecology are given among others by Aufhauser and Fischer, 1984; Fienberg and Meyer, 1983; Imrey et al., 1982; Whittam and Siegel-Causey, 1981. Whittam and Siegel-Causey (1981) for example, specified and selected a set of log-linear models to estimate species interactions and community structure for seabird colonies in Alaska.

Each of the estimation procedures can now be operationalized by a standard computer program. In case of log-linear models and linear logit models the GLIM (Generalized Linear Interactive Modelling) package developed by The Royal Statistical Society and NAG (Numerical Algorithms Group) makes use of the iterative WLS procedure (see Baker and Nelder, 1978 for a release manual of the GLIM-3 version and Gilchrist, 1982 for a discussion of the extended GLIM-4 version). Some qualitative statistical models based on survey data are analysed by Bowlby and Silk (1982) by making use of the GLIM package. The GENCAT package (Landis et al., 1976) provides a non- iterative WLS estimation of the logit models appropriate for the problems in cell (f) from Table 5.1.

The goodness-of-fit of any model within the GLM approach can be assessed when a model has been estimated. The goodness-of-fit statistic, also called deviance, is computed automatically within the GLIM package and takes the form:

$$D = -2 \left[\log_e \Lambda(\underset{\sim}{\beta}) - \log_e \Lambda_{max} \right],$$ (5.40)

where $\log_e \Lambda(\underset{\sim}{\beta})$ denotes the maximized log-likelihood of the fitted model and $\log_e \Lambda_{MAX}$ denotes the maximized log-likelihood of the 'complete' model, reducing the residual variation to zero (see also Nelder, 1974).

In case of logistic/logit models, the 'deviance' specializes to the form $-2 \log_e \Lambda(\underset{\sim}{\beta})$ (i.e. -2 times the maximized log-likelihood value of the fitted model). In case of dichotomous logit models such as in equation (5.28) the deviance becomes:

$$D = -2 \sum_{i=1}^{N} \left[y_i \log_e p_i + (1 - y_i) \log_e (1 - p_i) \right].$$ (5.41)

The deviance value can be determined in an analogous way for log-linear models, and it becomes in case of a three-dimensional contingency table:

$$\begin{aligned} D &= -2 \sum_{i=1}^{I} \sum_{j=1}^{J} \sum_{k=1}^{K} n_{ijk} \left[\log_e (m_{ijk}/n_{ijk}) \right] \\ &= 2 \sum_{i=1}^{I} \sum_{j=1}^{J} \sum_{k=1}^{K} n_{ijk} \left[\log_e (n_{ijk}/m_{ijk}) \right]. \end{aligned}$$ (5.42)

Expression (5.42) is in the literature of log-linear modelling usually referred to as G^2, and is called the likelihood-ratio statistic (see Bishop et al., 1975; Everitt, 1977).

An alternative type of goodness-of-fit statistic for logistic/logit and log-linear models is Pearson's chi-squared statistic, X^2 with:

$$X^2 = \sum_{i=1}^{I} \sum_{j=1}^{J} \sum_{k=1}^{K} \frac{(n_{ijk} - m_{ijk})^2}{m_{ijk}}$$ (5.43)

for log-linear models, and

$$x^2 = \sum_{i=1}^{N} \frac{(y_i - p_i)^2}{p_i \, (1 - p_i)} \tag{5.44}$$

for the logistic/logit models of expressions (5.28) and (5.31).

The deviance value and the chi-squared measures both have the same asymptotic χ^2 distributions with degrees of freedom equal to:

degrees of freedom = number of cell-elements in the contingency table

minus number of independent parameters in the model

that require estimating. (5.45)

The number of degrees of freedom is equal to the number of a priori restrictions. It can be shown that the likelihood-ratio measures have several important and useful features (e.g., divisibility into additive portions), which the chi-squared measures do not possess. The Pearson's x^2 is also proportional to the sample size and the probability of a type 2 error (acceptance of null-hypothesis when it is not true) becomes smaller for the statistic x^2 (see also Hagenaars, 1985). Fienberg (1980) mentioned a disadvantage on the G^2-statistic for small cell-frequencies, because it may underestimate the probability of type 1-errors (incorrect rejection of the null-hypothesis), and therefore it has a larger value than expected. For these reasons the deviance statistics are the most widely adopted goodness-of-fit measures. Model selection procedures for both log-linear and logistic/logit models are essentially based upon the goodness-of-fit measures described above. In log-linear modelling, for example, the G^2 statistics for various types of log-linear models are compared, and the most parsimonious member of the set which has a satisfactory fit to the observed data is the model which will be selected as the 'acceptable' representation of the structural relationships between the variables in the contingency table.

Systematic and efficient model selection procedures are required in case of multidimensional contingency tables where there are many possible models. The most widely adopted selection procedures are stepwise selection, screening and a simultaneous test procedure of parameters. Fingleton (1981) discussed these selection procedures in the frame of spatial dependence data:

(i)   stepwise selection, proposed by Goodman (1971). It is a stepwise procedure for the selection of log-linear models and includes both the forward selection and backward elimination procedure. The forward selection procedure adds significant effects to the basic model (e.g., the log-linear model with only the grand-mean effect included). The backward elimination procedure eliminates insignificant effects from the saturated model.

(ii)   screening (Brown, 1976). Each term in the saturated log-linear model is evaluated using two test-statistics, one of partial association and the other one of marginal association. Each term will be assessed in two extreme situations: the first conditional on all terms of the same order, and the second conditional on only the lower-order effects related to the term in question. They can be thought of as lower and upper bounds of the conditional G^2-values that would be obtained by adding that particular term to a previous specification of the log-linear model. As such, the two tests enable each term in the saturated model to be screened, and to be placed into one of three alternative categories:

(a) significant and necessary in the final 'acceptable' model;

(b) insignificant and unneccessary in the final model;

(c) of questionable significance and in need of further investigation.

Screening of parameters of the saturated log-linear model in this way will select an initial approximation to the final 'acceptable' model. The initial model may be refined by making use of a forward selection or backward elimination method.

(iii) simultaneous test procedure (Aitkin, 1979; 1980). A common characteristic of selection strategies for log-linear models is that they involve multiple tests of data. Some allowance should be made to compensate for the multivariate testing, to avoid attributing significance to what is merely random variation. Significance levels of statistical tests should be adjusted to compensate for the multiple testing. Significance levels are adjusted in the simultaneous test procedure and a systematic and internally consistent approach to model selection is provided.

Aitkin (1980) determined an overall type one error rate γ for the null hypothesis that a set of parameters are simultaneously equal to zero. Such a rate γ means the chance of at least one incorrect rejection of a true null hypothesis, and Aitkin recommends to select that level of insignificance α such that γ will be in the range between 0.25 and 0.50.

5.3.5. A graph approach to multidimensional contingency tables

A causal model interpretation of the statistical relationships between stimulus-response variables was discussed in subsection 5.2.4 in terms of path analysis. An extension of causal modelling for categorical data has been developed recently. A very appealing feature of hierarchical log-linear mod-els is that they have a minimal set of sufficient statistics which contain the relevant information for the joint distribution function. Unfortunately all variables are treated alike for most of the hierarchical log-linear mod-els. A modified path analysis approach was proposed by Goodman (1973) for qualitative variables when some variables are posterior to others. Goodman's analogue to conventional path analysis has two features. First, the models in the system make explicit the assumed order of priority (or causal ordering) of the categorical variables. Second, the relationships between the variables are denoted in terms of path diagrams (Wermuth and Lauritzen, 1982; Wrigley, 1985).

However, unlike conventional path analysis (discussed in subsection 5.2.4), there is no simple method of assigning numerical values (i.e., path coeffi-cients) to the paths and arrows which connect together the causal structure or path diagram.

Directed graphs with vertices and edges represent a system of dependencies which can be interpreted as a system of causal relationships. The recursive models of dependencies for qualitative variables are specified as a subclass of the models proposed by Goodman (see also Wermuth and Lauritzen, 1982). An example of a graph representation of a log-linear model is presented, which is denoted by three vertices and no response variable in Figure 5.4a, one response variable in Figure 5.4b and two response variables in Figure 5.4c. An example of an exploratory analysis with a causal ordering of categorical

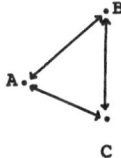

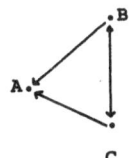

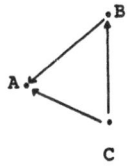

Figure 5.4(a).
A recursive system
with no response
variable.

Figure 5.4(b).
A recursive system
when A is a
response variable.

Figure 5.4(c).
A recursive system
when A and B are
response variables.

variables is presented by Pannekoek and Stronkhorst (1981). The study deals with the choice of a secondary school for boys and girls and the way that choice is affected by their socio-economic status, teacher's recommendation and also by the student's achievement level.

5.4. MULTIVARIATE ANALYSIS FOR NON-METRIC INFORMATION

5.4.1. Introduction

The phenomena, which are linked to each other in an IEM, and which are measured at either a nominal, an ordinal, an interval or a ratio scale, are multivariate in nature , and may be complexly structured. The multivariate phenomena can be based on spatially disaggregated regional profiles, such as economic profiles (e.g., labour supply/demand, industrial productivity), environmental profiles (e.g., recreational attractiveness), or infrastructural profiles (e.g., accessibility by road or by water). The major purpose in such case can be described as finding a certain structure in such a set of multivariate observations with various phenomena. Multivariate analysis deals with models which consist of more than one variate, with a variate defined to be a quantity which may take any value within a specified set like categories or rank numbers.

Two scaling methods for qualitative data (nominal and ordinal) will be discussed in this section, viz. multidimensional scaling (MDS) in the next subsection, and homogeneous scaling (especially HOMALS, or Homogeneity Analysis by Alternating Least Squares) in subsection 5.4.3. Both scaling procedures aim at deriving a quantitative configuration of a system which is measured on a qualitative scale. A goodness-of-fit criterion is introduced here to determine the fit between the qualitative data input and the quantitative or metric configuration.

The purpose of an MDS-procedure is to find a configuration of points with a ranking of the distances between these points which represent, as closely as possible, the ranking from the original data set with an ordinal scale of measurement. An MDS-procedure is especially relevant in case of ordinal information, with a ranked ordering between the values of a phenomenon. This will be discussed in subsection 5.4.2. A homogeneous scaling procedure such as HOMALS is developed for discrete or nominal variables, with individuals or objects being classified according to a limited number of categories or classes. This will be discussed in subsection 5.4.3.

Multidimensional scaling procedures and homogeneous scaling procedures have originally been developed in psychometrics, with an experiment ranked or classified by objects or subjects according to some point of view. However, both approaches also have shown their relevance in urban policy modelling and environmental policy modelling. An example of a simultaneous consideration of economic, social, physical and environmental profiles in a spatial system is presented by Blommestein and Nijkamp (1984) and by Nijkamp and Voogd (1984).

A multivariate analysis of water quality control involved in pollution and eutrophication in aquatic environments is presented by Ikeda and Itakura (1983).

5.4.2. Multidimensional scaling for ordinal information

The rationale behind the use of MDS is to transform ordinal data into cardinal units (or metric information); the MDS approach is able to take into account nonlinearities from the underlying relationships in a data matrix. Ordinal multidimensional scaling is sometimes also called ordinal geometric scaling. It takes for granted that a set of N objects or indicators in a K-dimensional space are transformed into and represented by cardinal information of a lower dimensionality, say S. The aspects or the attributes of the phenomena concerned are depicted in a S-dimensional Euclidean space in such a way that it should be consistent with the observed ordinal rankings. The attributes which are based on ordinal information, are represented as points in an Euclidean space. The configuration depicted in a S-dimensional Euclidean space with interpoint distances between the S-dimensional coordinates of the attributes needs to reflect a maximal agreement with the original ranking of attributes. The MDS approach then uses the degrees of freedom to obtain a metric representation of data which are originally ordinal in nature.

Suppose a symmetric NxN paired comparison table Δ with similarities and dissimilarities between N objects being expressed by ordinal numbers. The elements δ_{nm} of Δ represent ordinal rank numbers for the (dis)similarity between points n and m ($n,m = 1,..,N$). Each object n (which may be an indicator or a region) can be characterized in a S-dimensional space by the MDS-procedure with coordinates x_{ns} ($s = 1, ..., S$). The Euclidean distance d_{nm} between each pair of points (x_{ns},x_{ms}) is:

$$d_{nm} = (\sum_{s=1}^{S} (x_{ns} - x_{ms})^2)^{\frac{1}{2}} , \quad n, m = 1, ..., N. \tag{5.46}$$

The distance metric in equation (5.46) is symmetric and the triangle inequality holds. Besides, it is invariant for translation and homogeneous in nature.

The aim of the analysis now is to assess the coordinates x_{ns} and x_{ms} in such a way that the distance d_{nm} is in agreement with the original rankings δ_{nm}, so that such a distance has a maximal goodness-of-fit with respect to the original data. This condition implies that a stress function (or loss function) has to be minimized. The stress function will minimize the residual variance between all distances d_{nm} and δ_{nm}. However, the dissimilarity measure δ_{nm} is ordinal in nature and arithmetic operations are not permitted on ordinal information as was already mentioned before. A metric auxiliary variable d_{nm} is calculated for that reason which is in agreement with δ_{nm}, so that $d_{nm} < d_{nl}$ whenever $\delta_{nm} < \delta_{nl}$ (this is called the condition of order-isomorphism). Then the following goodness-of-fit function may be specified as a Minkowski-metric:

$$\phi = - \left[\frac{\sum\limits_n \sum\limits_m (d_{nm} - \hat{d}_{nm})^2}{\sum\limits_n \sum\limits_m d_{nm}^2} \right]^{\frac{1}{2}} \qquad (n \neq m) \qquad (5.47)$$

with the sum of squared distances as the normalization constant. The smaller the value of the loss function, the better the order-isomorphism (or monotonicity) between the original rankings, and the interpoint distances defined in equation (5.46). A set of coordinates x_{nm} is then found in an iterative procedure in which the loss function is minimized.

The above mentioned steps of the MDS procedure are summarized in Figure 5.5 below.

The analytical representation of the MDS approach can be formalized in the following way:

$$\min_x \quad \phi = f(D-\hat{D}) \qquad (5.48)$$
$$\text{subject to:} \quad \hat{D} \stackrel{m}{=} D$$
$$D = g(x),$$

where $\hat{D}$ and D are both distance matrices of order NxN with elements $\hat{d}_{nm}$ and d_{nm} respectively, and where $\underline{m}$ denotes the condition of order- isomorphism. An example of a widely used distance function $g(x)$ is already defined in (5.46).

The basic concepts of MDS-procedures are developed by Kruskal in the early sixties and are extensively discussed by Kruskal and Wish (1978). The MDS-procedure has been applied in various fields. Applications of MDS- procedures in policy modelling are given by Ikeda and Itakura (1983) for water quality modelling, Adelman and Morris (1974) for the measurement of levels of devel

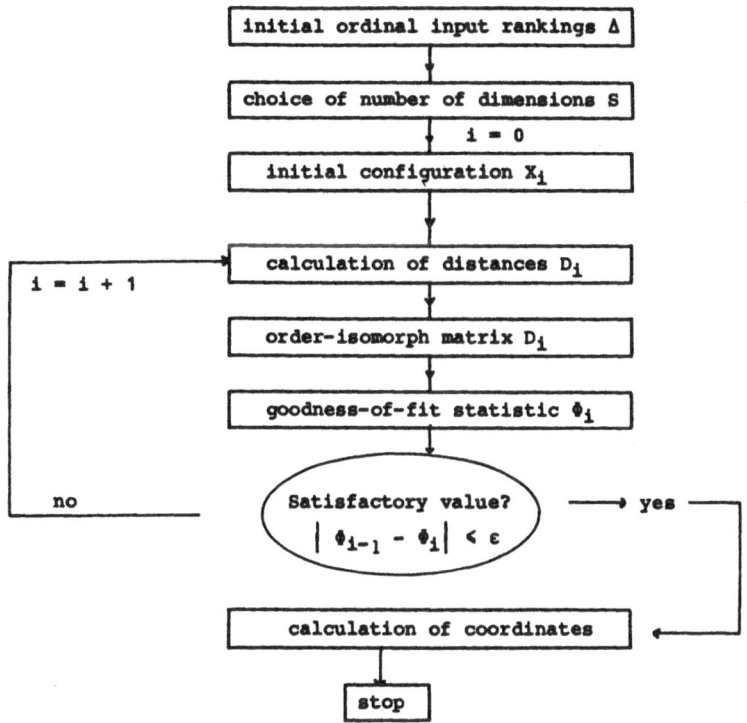

Figure 5.5. Simplified flow chart representation of the MDS procedure.

opment from low income nations, Nijkamp (1982) for an analysis of regional income determinants with ordinal and metric information in econometric models, and Nijkamp and Voogd (1984) for an analysis of Dutch regional infrastructure profiles.

5.5.3. Homogeneous scaling for nominal information

The aim of homogeneous scaling procedures for nominal information is to identify the relationships between a set of, say M, qualitative variables. Each variable has a number of categories for each of N individuals or objects, and each of the M variables has scores $1,..,k_j$ $(j=1,..,M)$.

Nominal or ordinal answer categories for questions from a survey and its respondents are represented into a joint Euclidean space in such a way that each respondent lies as close as possible to his/her answer categories. HOMALS (Homogeneity Analysis by Alternating Least Squares) is a program for homogeneity analysis; the program makes use of an alternating least squares algorithm (see also Gifi, 1981; Molenaar, 1985). It is an alternating procedure with person and category coordinates which are alternatingly improved.

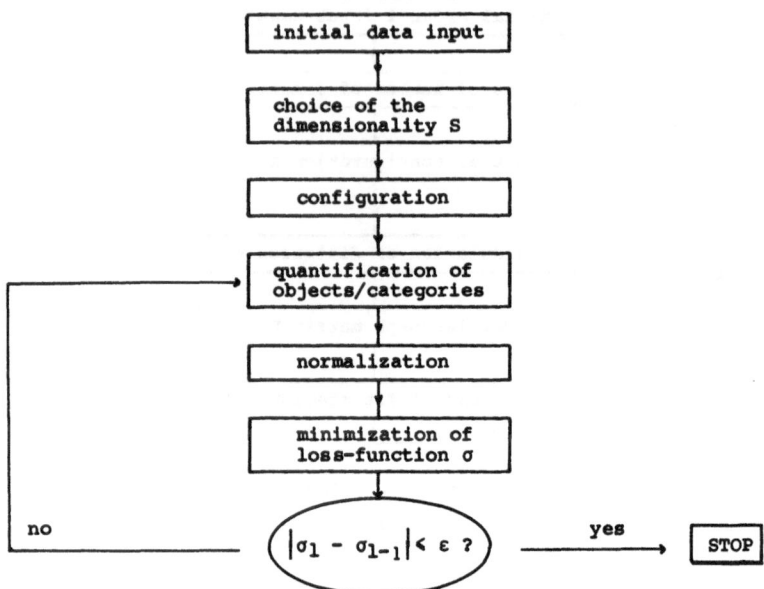

Figure 5.6. Representation of a HOMALS procedure.

All object and category scores are represented in a S-dimensional Euclidean space in such a way that the average squared distance between category points and the corresponding object points are minimized. The main characteristics of the iterative structure of the HOMALS program are represented in Figure 5.6.

In HOMALS the categories of nominal variables are quantified in such a way that the largest eigenvalue from the corresponding correlation matrix is maximized. This maximization of homogeneity will be achieved by means of a minimization of a loss function. Hence HOMALS deals with the correlation between each pair of variables by using a scaling procedure with respect to the initially qualitative labelling of the categories.

Examples of the application of the HOMALS computer program are given by Coolen and Van Rijckevorsel (1981) for an analysis of Dutch law and order mentality during the seventies, Israëls (1981) for an analysis of national statistics of shoplifting in the Netherlands, and Blommestein and Nijkamp (1984) for an analysis of qualitative relations on natural resources.

5.5. CONCLUDING REMARKS

The relevance of the three aspects discussed in Sections 5.2 to 5.4 will be discussed in this section in relation to the satellite concept of integration mentioned in Chapter 4. This will be summarized in Table 5.6 (see Brouwer and Nijkamp, 1986b).

The lack of reliable information would lead to scepticism regarding numerically quantified modelling results. When either theoretical knowledge concerning the specification of an IEM is scarce and/or when the relevant information is not reliable or not available at all, graph-theoretic methods and qualitative calculus are useful tools to analyse the structure of the impacts between variables or to solve a set of equations in a qualitative way.

Qualitative and quantative statistical models, which can be interpreted as members of a family of generalized linear models, are a useful tool in an IEM when different levels of measurement are available concerning the variables. The phenomena of an IEM which are linked to each other may be non-metric (nominal or ordinal) and multivariate in nature. Two scaling methods have been discussed, i.e. multidimensional scaling for ordinal information, and homogeneous scaling for nominal information. The originally non-metric information may be represented by and reduced into a quantitative configuration.

Table 5.6. Relevance of the methodological aspects in an IEM.

Methodological aspects	Relevance in an IEM
1. Structure analysis with binary or qualitative information	- lack of a sufficient data base - lack of reliable information - model interpretation with causality analysis and a minimum level of information
2. Different levels of measurement of the variables	- metric and non-metric information in the model operationalization - a coherent set of linear statistical models with the GLM approach - exploratory analysis for survey data with a nominal scale
3. Multivariate analysis for nominal and ordinal information	- representation and reduction of non-metric information in a quantitative configuration
4. Spatial scale and aggregation level	- variables with a spatial dimension and/or an allocation/diffusion between different regions

Finally the spatial scale level and spatial aggregation level both are relevant for all variables with a spatial dimension because regions are not uniquely demarcated entities, while various kinds of spatial interaction or
diffusion mechanisms may take place.

A satellite design of integration of an IEM is a hierarchical systems structure, with a core module representing the central aspects from the analysis.
The causal model structure can be analysed in case of zero/one information
concerning the impacts to trace out interdependencies and the hierarchical
levels. Additional information with respect to the signs of the impacts between variables may be used to determine the impacts of policy instruments on
the response variables in qualitative terms.

The relevance of the satellite design of integration as well as of the tools
presented in this chapter will be discussed in the next chapters within the
framework of an integrated environmental modelling approach for the Biesbosch
area.

PART C:

AN INTEGRATED ENVIRONMENTAL MODELLING APPROACH

CHAPTER 6. TOWARDS AN INTEGRATED ENVIRONMENTAL MODEL FOR THE BIESBOSCH AREA

6.1. INTRODUCTION

The presentation in Chapter 4 concerning the discussion of systems theory showed that a modular approach, with a hierarchical representation of modules representing the key phenomena, is a useful methodological framework of integration to a systematic development of an IEM. The major tasks in the use of systems theory to develop an IEM are twofold, viz. the formulation of the design from a coherent system to analyse the scope of a real world phenomenon, and the definition from a set of mathematical models to analyse the phenomena in an adequate way. The various stages in the development of a mathematical model from a systems approach - in terms of the definition and evaluation of a systems model - were presented in Figure 4.4 from subsection 4.2.1. The stages 1 to 3 in that figure deal with the definition and structure of a system and its subsystems, as well as with the selection and treatment of key phenomena. These three stages will also be called the formulation of the design stage of a coherent system, and begin the process from the systems definition phase to the mathematical modelling phase.

The first issue to be treated in this chapter (Section 6.2) is the presentation and discussion of the design stage of an IEM for the Biesbosch area in the Netherlands.

The Biesbosch area is a region in the south-western part of the Netherlands with many recreational facilities associated with the rivers and streams (such as fishing, swimming, sailing, surfing) (see also Van der Ploeg et al., 1984). Such recreational activities may be relevant for the development of employment in the region. However, the recreational activities also may be conflicting with nature conservation. Within the framework of an integrated modelling approach, the Biesbosch area is then considered to be a system with sub-systems which are actually based upon recreational activities, (regional) economic development, ecological processes and demographic characteristics. A mutual dependence between recreational activities, (regional) economic development, ecological processes and demographic characteristics is included in the analysis, and the relationships are represented in terms of links between modules. Such modules will be interpreted (in accordance with the definition in Section 2.1) as a set of interrelated variables, which have their background in a specific identifiable part of a compound environmental phenomenon.

The second issue to be treated in this chapter, which will be discussed in Section 6.2, deals with the association between modules. The aggregated approach at the level of modules will be disaggregated to the level of variables. The systems phase concerning the selection and treatment of key phenomena will be used in terms of the satellite design of integration.

Section 5.2 showed that, in case of the analysis of the structure of an IEM, a distinction can be made between different levels of measurement, i.e., binary or zero/one information, qualitative information (+, -, or 0), and metric information. First, a systems model representation of an IEM can be interpreted on the basis of binary or zero/one information. Subsection 5.2.2 showed a graph representation of a systems model to determine dependencies in a model, i.e. the hierarchical ordering, causality and key driving forces of the variables. This causality approach will be discussed in Section 6.3 for the model presented in Section 6.2. The second approach concerning the different levels of measurement deals with the analysis of systems models when the impacts between variables are represented in a qualitative way by their signs. The sign-solvability approach will be discussed in Section 6.4 for a dynamic simulation model to determine the qualitative impacts of policy instruments. The hierarchical ordering of variables, derived from the causality model structure, will also be helpful in that section for a recursive procedure to solve a model with qualitative information.

It will be concluded that the use of binary information and qualitative information within the framework of integrated environmental modelling may provide practical tools to draw reliable conclusions concerning analytical and policy aspects.

The relevance of outdoor recreation with respect to nature conservation and the availability of recreational facilities will be further elaborated in the next chapter. Recreational activities in the Biesbosch area will be analysed in Chapter 7 with respect to, among others, the relevance of phenomena such as the nature of landscape and the availability of recreational facilities, and a cause-and-effect analysis between type of boat and type of recreational activities.

6.2. DESIGN OF AN INTEGRATED SYSTEMS MODEL AND THE TREATMENT OF KEY PHENOMENA

Three designs for integrating environmental phenomena will be discussed in this section in order to define and structure a system with its sub-systems, as well as to define the links between the modules of an IEM mentioned in Section 6.1. Such links between the modules depict the association between variables at an aggregated level. A distinction will be made here between a

horizontal, a vertical and a satellite design structure (see also the discussion of these designs for model integration in subsection 4.3.2 and Section 4.4).

A schematic representation of the links between modules is presented in Figure 6.1 in terms of a horizontal framework of integration. A horizontal
model concept of integration (in terms of a parallel design), like the one
presented in Figure 6.1 is characterized by equal position and contribution
of the separate modules. However, in this case an essentially hierarchical
representation of the links between modules would be desirable here in order
to express the recreational activities in the area as the key factor or dri-

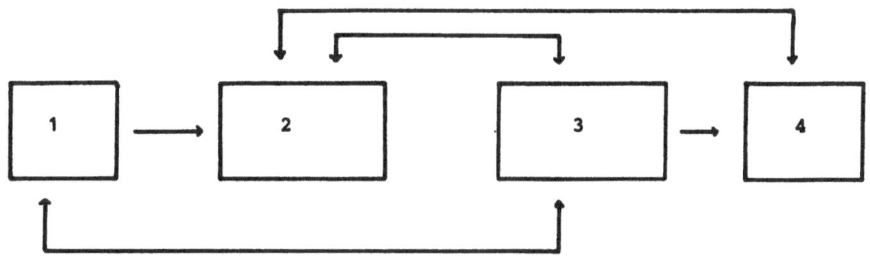

1= demography module;   3= regional-economy module;
2= recreation module;   4= natural-environment module.

Figure 6.1. Outline of a horizontal model design of integration.

ving force for other modules. Recreational activities are regarded as the
central research issue of the study of the Biesbosch area, with a mutual
dependence between a natural environment module and a regional-economic module. A presentation of such a so-called hierarchical (or vertical) model
concept of integration is found in Figure 6.2. The vertical model approach
from Figure 6.2 may be useful, for example, in case of a separate analysis of
economic aspects related to recreational activities or of environmental aspects related to recreational activities.

A hierarchical ordering or vertical structure of phenomena is also a fundamental concept in systems theory (see also Simon, 1973, who interpreted complex systems as an organization of hierarchies). A vertical model approach
which focuses on preselected topics, can be chosen for pragmatic and computational reasons (for instance, because a horizontal model approach may be
overambitious in terms of staff time and necessity of information) (see also
subsection 4.3.2). The hierarchical model structure in Figure 6.2 is based on
two levels, viz. the first level with the recreation module, and the lower
level which is made up by the three remaining modules and their interrelationships (natural environment, regional economy, and demography). An example

of a hierarchical model approach in the survey from Chapter 2 was presented
in Section 2.4. This model is concerned with a policy analysis of water management in the Netherlands, and included a water distribution module which
simulates and regulates the balance between the supply and demand of water
from the different sectors.

Unfortunately, a direct relationship between the regional-economic module and
the natural environment module (e.g. the relationship between the natural
environment and the employment affiliated with recreational activities) will
be analysed at best marginally in the structure of a hierarchical (or vertical) design of integration in Figure 6.2.

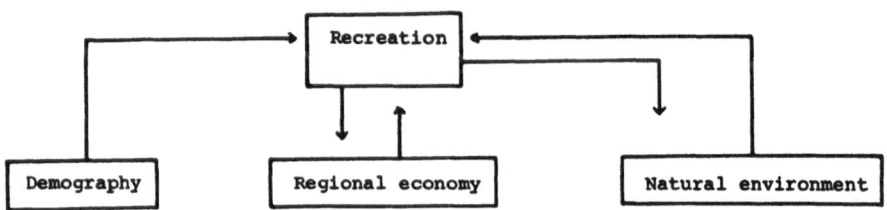

Figure 6.2. Outline of a vertical model design of integration.

The major emphasis in Figure 6.2 is thus placed upon the (partial) dominance
relationships between the core module and the separate modules, while simul-
taneity of relationships (interactions) between the modules is neglected.

Finally, an outline of the satellite concept of integration is presented in
Figure 6.3 (see also Brouwer and Nijkamp, 1986b).

The development of the satellite concept of integration towards an IEM for
the Biesbosch area consists of three steps:

(i)   the recreational aspects of each activity sector (e.g., regional econo-
       my) are identified. This implies the determination of the recreational
       aspects related to the regional-economic module, the natural environ-
       ment module, and the demographic module respectively. The recreational
       aspects of the regional-economic module may include, for example, the
       amount of money spent by recreationers for food and the maintenance of
       boats; the recreational aspects of the natural environment module may
       determine the level of the set foot on different types of shores in the
       Biesbosch area. The recreational aspects related to the demographic
       module may be spatially classified in terms of the demographic nature
       of recreationers (e.g., their age or their level of education);

(ii)   the recreational aspects from the separate modules are combined in one
       integrated recreational module, which is the core module of the analy-
       sis;

(iii) the other aspects of the analysis of the separate modules are operationalized in the final step. For instance, the mutual dependence between
the regional-economic and the demographic module may include a balance
between the supply of and the demand for employment in the Biesbosch
area, depending on the level of recreational activities as well as on
other types of employment. A change in supply of recreational facilities (which is part of the regional-economic module) may influence
vegetation or types of banks (which is part of the natural environmental module).

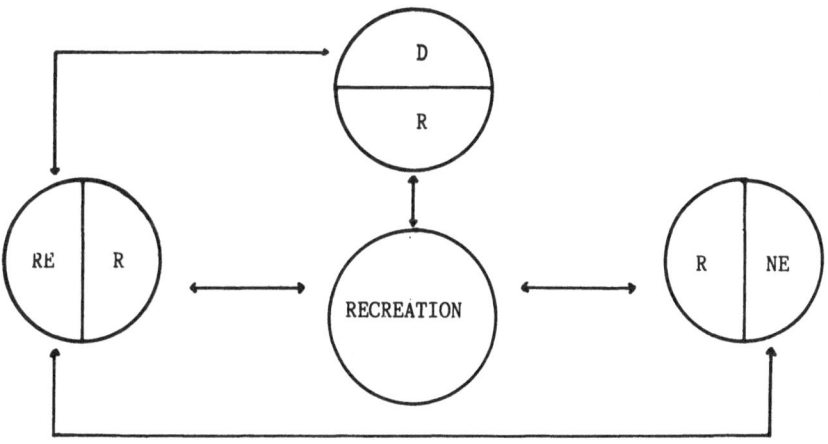

R = Recreation;   RE = Regional economy;
D = Demography;   NE = Natural environment.

Figure 6.3. Outline of the satellite model design of integration.

The monodisciplinary relationships are covered in steps (i) and (ii), while
the relationships which cross the boundary of a single discipline are treated
in step (iii). The first stage of the definition and evaluation of a systems
model for the Biesbosch area deals with the definition and structure of a
system, according to stage 1 from Figure 4.4.
A systems model is defined in Figure 6.3 with four modules of the systems
concerned. The structure of the systems model is outlined in terms of the satellite design of integration: recreational activities are the key phenomena
of the model.
The hierarchical model structure (in accordance with the systems hierarchy
discussed in Sections 4.2 and 4.4), with interdependent subsystems within the
satellite concept of integration is based upon three levels, viz.:

(i)   the first level is the core of the analysis which represents the re-creation module;

(ii)   the second level consists of the aspects of the core module which are related to the other modules (i.e., the recreational aspects with res-pect to regional economy, natural environment, and demography);

(iii) the third level is based upon the lower level modules and their inter-relationships.

Having formulated the satellite structure at the level of modules, which is the design stage for a coherent system in order to analyse the scope of a real world phenomenon, the relationships between the individual variables will now be formulated. This phase of model operationalization deals with the selection and treatment of key phenomena and the selection of the relation-ships (see stages 2 and 3 from Figure 4.4). Figure 6.4 below shows the links between the successive variables.

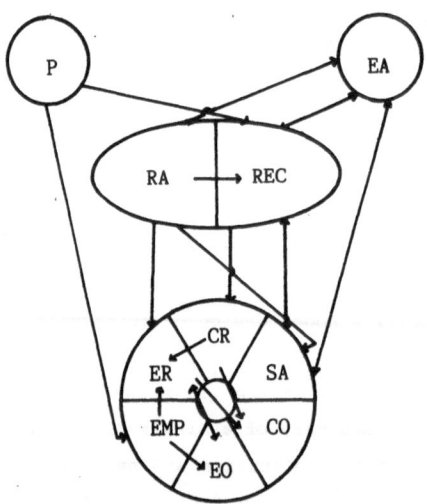

Figure 6.4. Outline of an IEM at the level of variables.

The four modules from Figure 6.3 are disaggregated at the level of variables in Figure 6.4.

The 10 variables from Figure 6.4 are interpreted in the following way:

RA : recreational attractiveness (a supply variable concerning the availability of recreational facilities);

P : population size;

REC : demand for recreational facilities;

EA : environmental attractiveness;

SA : stimulating regional activities;

EMP : supply of labour;

CR : expenditures on consumption goods by recreationers;

CO : non-consumptive expenditures by recreationers;

ER : demand for employment, related to recreational activities;

EO : demand for employment, related to 'other' activities (i.e. not directly
related to recreational activities).

The variables RA and P are exogenous in nature (also called stimulus variables), and the other variables are endogenous variables (or called response
variables).

The recreation module is composed of the variables RA and REC; the demography
module and the natural environment module are composed of respectively the
variables P and EA, and the (regional) economy module is composed of the
variables SA, EMP, CR, CO, ER and EO.

The variable REC which is part of the recreation module is the core variable
concerning the impacts between variables, because it either stimulates the
level of variables in the other modules, or it is the response to the level
of variables in other modules.

One of the conclusions from Chapter 3 concerning the evaluation of the IEMs
from the survey was that appropriate information and adequate experiments in
modelling environmental phenomena and processes are rather scarce, because of
the complexity and variability of environmental systems. Table 3.2, for example, mentioned that the use of simulation models is, in such cases, a widely
used and relevant tool in modelling environmental processes.

A dynamic simulation model will be presented below in terms of linear difference equations for the variables discussed above from the four mentioned
modules. It has been developed on the subject of modelling environmental
phenomena due to scarcity of reliable information and adequate experiments
concerning the variables.

In the sequel of this chapter the relationships between variables from Figure
6.4 will be analysed in various ways, depending on the precision of information on the impacts.

A dynamic simulation model of an IEM will be presented below in equation
6.1, followed by a discussion of the individual equations.

The model represents an assessment of the impacts exerted by two exogenous variables in long-term regional development. The exogenous variables, which are an input for the endogenous variables, are the recreational attractiveness (RA) and the size of population (P). The eight equations from (6.1) will be explained briefly below, and they are presented in terms of rates of change for variables between period t and period t-1.

(i) $\quad SA_t = SA_{t-1} + \alpha_1 EA_t - \alpha_2 (REC_{t-1} - REC_{t-2}) + \alpha_3 RA_t$

(ii) $\quad EA_t = EA_{t-1} + \beta_1 SA_t - \beta_2 (REC_t - REC_{t-1}) - \beta_3 (REC_{t-1} - REC_{t-2}) + \beta_4 RA_t$

(iii) $\quad REC_t = REC_{t-1} + \gamma_1 (SA_t - SA_{t-1}) + \gamma_2 (EA_t - EA_{t-1})$
$\qquad\qquad - \gamma_3 (EA_{t-1} - EA_{t-2}) + \gamma_4 RA_t + \gamma_5 P_t$

(iv) $\quad EMP_t = EMP_{t-1} + \delta_1 P_t$ (6.1)

(v) $\quad CR_t = CR_{t-1} + \epsilon_1 (REC_t - REC_{t-1})$

(vi) $\quad CO_t = CO_{t-1} + \eta_1 (REC_t - REC_{t-1}) + \eta_2 (CR_t - CR_{t-1})$

(vii) $\quad ER_t = ER_{t-1} + \lambda_1 (REC_t - REC_{t-1}) + \lambda_2 (EMP_t - EMP_{t-1}) + \lambda_3 (CR_t - CR_{t-1})$
$\qquad\qquad + \lambda_4 (CO_t - CO_{t-1}) + \lambda_5 (EO_t - EO_{t-1}) + \lambda_6 RA_t$

(viii) $EO_t = EO_{t-1} + \mu_1 (EMP_t - EMP_{t-1}) + \mu_2 (ER_t - ER_{t-1})$

The rate of change in the development of regional activities will be stimulated by means of an increase of the level of environmental attractiveness as well as the recreational attractiveness, and it will be discouraged by an increase in demand for recreational activities (equation (i) in (6.1)).

The rate of change in the development of environmental attractiveness (in equation (ii)) is determined by the level of stimulating efforts for regional activities, the rate of change of the demand for recreational facilities, and the level of recreational attractiveness.

The rate of change in the demand for recreational facilities in equation (iii) is related to the rates of change in stimulation of regional activities and environmental attractiveness, as well as to the levels of recreational attractiveness and population.

The rate of change in the supply of labour force is determined by the population size (in equation (iv)).

Equation (v) denotes the rate of change in the expenditure for consumption goods by recreationers, and that variable is determined by the rate of change of the demand for recreational facilities.

The rate of change concerning the non-consumptive expenditures by recreationers (in equation (vi)) is a function of the rates of change in the demand for
recreational facilities and the expenditure for consumption goods by recreationers.

The demand for employment, according to the two types of recreational activities is described by the equations (vii) and (viii).

The rate of change in the demand for employment related to recreational activities (in equation (vii)) is determined by the rates of change of demand for
recreational facilities, supply of labour, and expenditures by recreationers
for consumption goods respectively, as well as non-consumptive goods, and the
level of recreational attractiveness.

Finally, the demand for employment related to 'other' activities in equation
(viii) is a function of the rates of change of the supply of labour and the
demand for employment related to recreational activities.

A matrix representation of the dynamic simulation model (6.1) with first-order and second-order time lags is given below in model (6.2).

$$
\begin{bmatrix}
1 & -\alpha_1 & 0 & 0 & 0 & 0 & 0 & 0 \\
-\beta_1 & 1 & \beta_2 & 0 & 0 & 0 & 0 & 0 \\
-\gamma_1 & -\gamma_2 & 1 & 0 & 0 & 0 & 0 & 0 \\
0 & 0 & 0 & 1 & 0 & 0 & 0 & 0 \\
0 & 0 & -\varepsilon_1 & 0 & 1 & 0 & 0 & 0 \\
0 & 0 & -\eta_1 & 0 & -\eta_2 & 1 & 0 & 0 \\
0 & 0 & -\lambda_1 & -\lambda_2 & -\lambda_3 & -\lambda_4 & 1 & -\lambda_5 \\
0 & 0 & 0 & -\mu_1 & 0 & 0 & -\mu_2 & 1
\end{bmatrix}
\begin{bmatrix}
SA \\ EA \\ REC \\ EMP \\ CR \\ CO \\ ER \\ EO
\end{bmatrix}_t
=
\begin{bmatrix}
1 & 0 & -\alpha_2 & 0 & 0 & 0 & 0 & 0 \\
0 & 1 & \beta_2 - \beta_3 & 0 & 0 & 0 & 0 & 0 \\
-\gamma_1 & -\gamma_2 - \gamma_3 & 1 & 0 & 0 & 0 & 0 & 0 \\
0 & 0 & 0 & 1 & 0 & 0 & 0 & 0 \\
0 & 0 & -\varepsilon_1 & 0 & 1 & 0 & 0 & 0 \\
0 & 0 & -\eta_1 & 0 & -\eta_2 & 1 & 0 & 0 \\
0 & 0 & -\lambda_1 & -\lambda_2 & -\lambda_3 & -\lambda_4 & 1 & -\lambda_5 \\
0 & 0 & 0 & -\mu_1 & 0 & 0 & -\mu_2 & 1
\end{bmatrix}
\begin{bmatrix}
SA \\ EA \\ REC \\ EMP \\ CR \\ CO \\ ER \\ EO
\end{bmatrix}_{t-1}
+
$$

$$
\begin{bmatrix}
0 & 0 & \alpha_2 & 0 & 0 & 0 & 0 & 0 \\
0 & 0 & \beta_3 & 0 & 0 & 0 & 0 & 0 \\
0 & \gamma_3 & 0 & 0 & 0 & 0 & 0 & 0 \\
0 & 0 & 0 & 0 & 0 & 0 & 0 & 0 \\
0 & 0 & 0 & 0 & 0 & 0 & 0 & 0 \\
0 & 0 & 0 & 0 & 0 & 0 & 0 & 0 \\
0 & 0 & 0 & 0 & 0 & 0 & 0 & 0 \\
0 & 0 & 0 & 0 & 0 & 0 & 0 & 0
\end{bmatrix}
\begin{bmatrix}
SA \\ EA \\ REC \\ EMP \\ CR \\ CO \\ ER \\ EO
\end{bmatrix}_{t-2}
+
\begin{bmatrix}
\alpha_3 & 0 \\
\beta_4 & 0 \\
\gamma_3 & \gamma_4 \\
0 & \delta_1 \\
0 & 0 \\
0 & 0 \\
\lambda_6 & 0 \\
0 & 0
\end{bmatrix}
\begin{bmatrix}
RA_t \\ P_t
\end{bmatrix}
\qquad (6.2)
$$

Integrated simulation models have been developed to identify the impacts of
development alternatives upon social, economic and environmental systems.
Some applications concerning the socio-economic and environmental consequences of development alternatives by means of simulation approaches were discussed in Sections 2.3, 2.7, 2.11, 2.12 and 2.14. The main relevance of integrated simulation models concerns the ability of such models to represent the
relationships between economic, demographic, social and environmental phenomena in a systematic way (see also Lonergan, 1983 who presented a simulation/
optimization model for natural resource planning).

The causal model structure of the stimulus-response relationships of an integrated system reproduced in equation (6.2) will be analysed in the following section. The causal systems structure will be analysed for a set of stimulus-response relationships to determine whether interdependencies exist between variables and to examine out the hierarchical ordering of the variables, when only binary information is available.

6.3. CAUSALITY ANALYSIS OF AN INTEGRATED ENVIRONMENTAL MODEL

The causal links of an IEM for the Biesbosch area at the level of variables were outlined in Figure 6.4 by means of stimulus-response relationships, and such links reflect the hierarchical nature of the model in terms of dependencies and interdependencies. The relevance of the use of graph theory for the interpretation of systems models when only binary information is available - such as the outline of an IEM in Figure 6.4 - was discussed in subsection 5.2.2.

The graph representation of the stimulus-response relations of the ten variables depicted in Figure 6.4 may be denoted by the adjacency matrix A with elements a_{ij} $(i,j=1,\ldots,10)$ which is defined as:

$$a_{ij} = \begin{cases} 1 \text{ if an edge exists from vertex } j \text{ into vertex } i \ (i,j=1,\ldots,10) \\ 0 \text{ if otherwise.} \end{cases}$$

The adjacency matrix of a graph G, denoted by A(G), is presented below. The empty cell-elements from the matrix are zero.

A representation of the causal model structure, which is based on the same definition as the adjacency matrix, is a graph G. Such a graph is defined by a set of vertices $V = (v_1, \ldots, v_{10})$ and a set of edges $E = (e_1, \ldots, e_{21})$, and is depicted in Figure 6.5 (a). The elements 1 to 10 in the figure are the

	RA	P	SA	EA	REC	EMP	CR	CO	ER	EO
1 = RA										
2 = P										
3 = SA	1			1	1		-			
4 = EA	1		1		1					
5 = REC	1	1	1	1						
6 = EMP		1								
7 = CR					1					
8 = CO					1		1			
9 = ER	1				1	1	1	1		1
10 = EO						1			1	

$A(G) =$ (shown in the matrix above)

vertices which belong to the set V, and the edges ((1,3), (1,4), ..., (10,9)) with 22 elements belong to the set E. The numbers in Figure 6.5 (a) corres- pond to the variables from the adjacency matrix A(G).

Figure 6.5(a) shows two sets of interdependent variables, viz. one set with the variables (SA, EA, REC) and another set with the variables (ER, EO). These two sets of variables are also called the strong components of the graph, because the variables from such a set are mutually dependent, and they correspond to an maximally connected directed subgraph. The graph where all these interdependencies are deleted but represented in terms of blocks, is called a reduced graph.

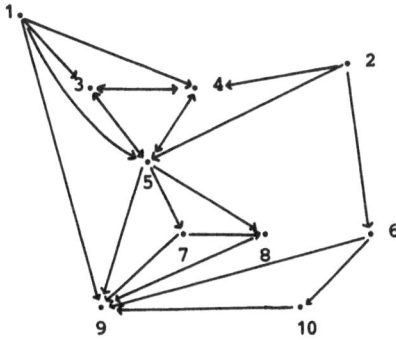

Figure 6.5(a) Graph-representation of the stimulus- response relationships G = (V,E).

Figure 6.5(b) Reduced-graph represen- tation of the stimulus- response relationships.

The reduced graph which is derived from the graph in Figure 6.5 (a) is pre- sented in Figure 6.5 (b). The properties derived from the reduced graph are useful for a hierarchical model representation, and the properties derived depend only on the causal links between variables. A reduced graph denotes the causal links - in terms of recursive model characteristics - between blocks of variables.

The recursive causal model structure will be analysed now by means of a top- down or hierarchical approach (Gilli, 1984). It is an iterative procedure to discriminate between the causal links, which relate blocks of variables with each other. An initial stage of the top-down approach is the reduced graph representation, and the exogenous variables are excluded from the reduced graph in the first iteration. The result from this exclusion process leads to a new set of stimulus-response relationships between the remaining variables, and this stage is called the first level. The first level becomes input for the second iteration, etc. The succeeding levels are obtained by variables which exclusively control the remaining variables, and this phenomenon is the reason that the hierarchical approach will also be called a top-down hierarchy.

Iteration 1: Construction of the adjacency matrix $A(G_R)$ of the reduced graph in Figure 6.5 (b). The indegree from a vertex is defined by the number of edges coming into the vertex. The adjacency matrix becomes in that case:

$$
A(G_R) = \quad
\begin{array}{c}
1 \\
2 \\
3,4,5 \\
6 \\
7 \\
8 \\
9,10
\end{array}
\begin{bmatrix}
0 & 0 & 0 & 0 & 0 & 0 & 0 \\
0 & 0 & 0 & 0 & 0 & 0 & 0 \\
1 & 1 & 0 & 0 & 0 & 0 & 0 \\
0 & 1 & 0 & 0 & 0 & 0 & 0 \\
0 & 0 & 1 & 0 & 0 & 0 & 0 \\
0 & 0 & 1 & 0 & 1 & 0 & 0 \\
1 & 0 & 1 & 1 & 1 & 1 & 0
\end{bmatrix}
\begin{array}{c}
\text{indegree} \\
0 \\
0 \\
2 \\
1 \\
1 \\
2 \\
5
\end{array}
$$

Level 1: The variables 1 and 2 are excluded in the first iteration of the top-down hierarchy, because they exclusively control the other variables, and have indegree equal to zero. The variables 1 and 2 are exogenous in nature for all remaining variables in the analysis, and they are therefore excluded in the first iteration.

Iteration 2: Construction of the reduced graph as well as the adjacency matrix concerning the remaining variables, when the variables 1 and 2 are excluded:

$$
A(G_R) = \quad
\begin{array}{c}
3,4,5 \\
6 \\
7 \\
8 \\
9,10
\end{array}
\begin{bmatrix}
0 & 0 & 0 & 0 & 0 \\
0 & 0 & 0 & 0 & 0 \\
1 & 0 & 0 & 0 & 0 \\
1 & 0 & 1 & 0 & 0 \\
1 & 1 & 1 & 1 & 0
\end{bmatrix}
\begin{array}{c}
\text{indegree} \\
0 \\
0 \\
1 \\
2 \\
4
\end{array}
$$

Level 2: The variables (3,4,5) and 6 have a zero indegree, which means that the other variables from iteration 2 will not influence these variables. The two sets with the variables (3,4,5) and 6 are excluded, because they are input to the remaining variables at this level.

<u>Iteration 3</u>: Construction of the reduced graph and adjacency matrix when the variables (3,4,5) and 6 from the previous level are excluded:

$$A(G_R) = \begin{array}{c} 7 \\ 8 \\ 9,10 \end{array} \left[\begin{array}{ccc} 0 & 0 & 0 \\ 1 & 0 & 0 \\ 1 & 1 & 0 \end{array} \right] \begin{array}{c} \underline{indegree} \\ 0 \\ 1 \\ 2 \end{array}$$

<u>Level 3</u>:   Variable 7 is input to the remaining variables from iteration 3. It will be excluded after iteration 3 because its indegree now becomes zero when all previous variables are excluded.

<u>Iteration 4</u>: Construction of the reduced graph and the adjacency matrix when variable 7 from the previous level is excluded:

$$A(G_R) = \begin{array}{c} 8 \\ 9,10 \end{array} \left[\begin{array}{cc} 0 & 0 \\ 1 & 0 \end{array} \right] \begin{array}{c} \underline{indegree} \\ 0 \\ 1 \end{array}$$

8.———→. (9,10)

<u>Level 4</u>:   Variable 8 will now be excluded and the set of variables (9,10) is the remaining one. The nature from this set of variables is determined by all other variables which are excluded in previous stages.

A summary of the stages is presented in Table 6.1.

Table 6.1. Top-down hierarchy of the variables from Figure 6.5.

Level	Variables excluded
1	RA, P
2	(SA, EA, REC), EMP
3	CR
4	CO
5	(ER, EO)

The hierarchical model structure concerning the stimulus-response relationships of the IEM from Figure 6.4 is shown by means of the top-down representation in Table 6.1, when moving from level 1 to level 5. The exogenous variables RA and P, which exclusively determine the nature of the other variables, are shown in the first level. The second level shows the variables which determine the nature and stimulus-response pattern of the remaining variables. This level shows that the variables SA, EA, REC and EMP are an input to the variables CR, CO, ER and EO; they stimulate the nature of lower level variables, without being affected by them.

Lower level variables from Table 6.1 are an input to the higher level variables and this phenomenon denotes the hierarchical model structure of the IEM outlined in Figure 6.4.

The variables RA and P, which are part of the recreation and the demography module respectively, are an input to the variables from the other modules. Three modules are mutually dependent, i.e. the environmental, the regional-economic and the recreation module are mutually related by means of the variables EA, SA, and REC, which are also input to the variables CR, CO, ER and EO in the economic module.

Having discussed now the causal model structure of the stimulus-response relationships from Figure 6.4, the systems model from (6.2) will be analysed in the following section when qualitative information is available.

6.4. ANALYSIS OF AN INTEGRATED ENVIRONMENTAL MODEL WITH QUALITATIVE INFORMA-
      TION

A causality representation of an IEM based on a systems model was discussed in the previous section when binary or zero/one relationships are the only information about the impacts between variables. A hierarchical representation of the impacts between variables was presented in Table 6.1. Graph theory is a useful tool in the case of zero/one information in order to trace interdependencies, as well as the various hierarchical levels of stimulus-response relationships. Additional information may also be available with respect to the signs of the impacts between variables. A qualitative approach is an appropriate way of dealing with a weak database, and it may be used to determine the impacts of policy instruments on response variables in terms of positive, negative or zero 'values'.

The relevance of the sign-solvability approach to deal with analytical or policy aspects of systems models was already discussed in subsection 5.2.3.

However, the conditions for sign-solvability with qualitative information are rather strict. This was recently fully explored in a dynamic simulation model of urban decline discussed by Brouwer and Nijkamp, 1986e, which appeared to be not sign-solvable when only qualitative information about the impacts was available. It was demonstrated however that a dynamic model may become - at least partially - sign-solvable when a mixture of qualitative and quantitative information is available.

The conditions of sign-solvability will be applied to the simulation model of an IEM, which was presented in Section 6.2. When the reduced form of the simulation model is sign-solvable, it will be possible to determine the qualitative impacts from the exogenous variables and the lagged endogenous variables on the endogenous variables.

Table 6.1 in the previous section summarized the hierarchical structure of the simulation model when moving from level 1 to level 5. The conditions of sign-solvability of lower level variables can be analysed independently of higher level variables, and this characteristic means that the conditions of sign-solvability can also be applied in this case to parts of the simulation model. The conditions of sign-solvability with purely qualitative information will now be further elaborated. The first condition of sign-solvability is that all diagonal elements of the impact matrix shall be negative, because that matrix will be inverted (see also subsection 5.2.3). This condition holds here when the signs in all columns of the matrices in (6.2) are reversed. Lancaster (1962) has shown that this type of sign reversements multiplies a variable by -1, and does not have any effect whether or not the system is sign-solvable.

The model in (6.2) then can be transformed into (6.3).

$$
\begin{bmatrix}
-1 & \alpha_1 & 0 & 0 & 0 & 0 & 0 & 0 \\
\beta_1 & -1 & -\beta_2 & 0 & 0 & 0 & 0 & 0 \\
\gamma_1 & \gamma_2 & -1 & 0 & 0 & 0 & 0 & 0 \\
0 & 0 & 0 & -1 & 0 & 0 & 0 & 0 \\
0 & 0 & \varepsilon_1 & 0 & -1 & 0 & 0 & 0 \\
0 & 0 & \eta_1 & 0 & \eta_2 & -1 & 0 & 0 \\
0 & 0 & \lambda_1 & \lambda_2 & \lambda_3 & \lambda_4 & -1 & \lambda_5 \\
0 & 0 & 0 & \mu_1 & 0 & 0 & \mu_2 & -1
\end{bmatrix}
\begin{bmatrix}
SA \\ EA \\ REC \\ EMP \\ CR \\ CO \\ ER \\ EO
\end{bmatrix}_t
=
\begin{bmatrix}
-1 & 0 & \alpha_2 & 0 & 0 & 0 & 0 & 0 \\
0 & -1 & -\beta_2+\beta_3 & 0 & 0 & 0 & 0 & 0 \\
\gamma_1 & \gamma_2+\gamma_3 & -1 & 0 & 0 & 0 & 0 & 0 \\
0 & 0 & 0 & -1 & 0 & 0 & 0 & 0 \\
0 & 0 & \varepsilon_1 & 0 & -1 & 0 & 0 & 0 \\
0 & 0 & \eta_1 & 0 & \eta_2 & -1 & 0 & 0 \\
0 & 0 & \lambda_1 & \lambda_2 & \lambda_3 & \lambda_4 & -1 & \lambda_5 \\
0 & 0 & 0 & \mu_1 & 0 & 0 & \mu_2 & -1
\end{bmatrix}
\begin{bmatrix}
SA \\ EA \\ REC \\ EMP \\ CR \\ CO \\ ER \\ EO
\end{bmatrix}_{t-1}
$$

$$
+
\begin{bmatrix}
0 & 0 & -\alpha_2 & 0 & 0 & 0 & 0 & 0 \\
0 & 0 & -\beta_3 & 0 & 0 & 0 & 0 & 0 \\
0 & -\gamma_3 & 0 & 0 & 0 & 0 & 0 & 0 \\
0 & 0 & 0 & 0 & 0 & 0 & 0 & 0 \\
0 & 0 & 0 & 0 & 0 & 0 & 0 & 0 \\
0 & 0 & 0 & 0 & 0 & 0 & 0 & 0 \\
0 & 0 & 0 & 0 & 0 & 0 & 0 & 0 \\
0 & 0 & 0 & 0 & 0 & 0 & 0 & 0
\end{bmatrix}
\begin{bmatrix}
SA \\ EA \\ REC \\ EMP \\ CR \\ CO \\ ER \\ EO
\end{bmatrix}_{t-2}
+
\begin{bmatrix}
-\alpha_3 & 0 \\
-\beta_4 & 0 \\
-\gamma_3-\gamma_4 & 0 \\
0 & -\delta_1 \\
0 & 0 \\
0 & 0 \\
-\lambda_6 & 0 \\
0 & 0
\end{bmatrix}
\begin{bmatrix}
RA_t \\ P_t
\end{bmatrix}
\qquad (6.3)
$$

Model (6.3) can in matrix terms be denoted by:

$$A y_t = B y_{t-1} + C y_{t-2} + D x_t \quad , \tag{6.4}$$

or it can also be represented into submatrices by means of a matrix decomposition as:

$$\begin{bmatrix} A_{11} & 0 \\ A_{21} & A_{22} \end{bmatrix} \begin{bmatrix} y_1 \\ y_2 \end{bmatrix}_t = \begin{bmatrix} B_{11} & 0 \\ B_{21} & B_{22} \end{bmatrix} \begin{bmatrix} y_1 \\ y_2 \end{bmatrix}_{t-1} + \begin{bmatrix} C_{11} & 0 \\ 0 & 0 \end{bmatrix} \begin{bmatrix} y_1 \\ y_2 \end{bmatrix}_{t-2} + \begin{bmatrix} d_{11} & d_{12} \\ d_{21} & d_{22} \end{bmatrix} \begin{bmatrix} RA_t \\ P_t \end{bmatrix} \tag{6.5}$$

The submatrices of the matrix A in model (6.5) can be denoted in qualitative terms by:

$$\text{sign } (A_{11}) = \begin{bmatrix} - & + & 0 & 0 \\ + & - & - & 0 \\ + & + & - & 0 \\ 0 & 0 & 0 & - \end{bmatrix}, \quad \text{sign } (A_{21}) = \begin{bmatrix} 0 & 0 & + & 0 \\ 0 & 0 & + & 0 \\ 0 & 0 & + & + \\ 0 & 0 & 0 & + \end{bmatrix}, \quad \text{sign } (A_{22}) = \begin{bmatrix} - & 0 & 0 & 0 \\ + & - & 0 & 0 \\ + & + & - & + \\ 0 & 0 & + & - \end{bmatrix}$$

Sign-solvability of the model in (6.4) means that the vector y_t can be expressed in qualitative terms as:

$$y_t = A^{-1} B y_{t-1} + A^{-1} C y_{t-2} + A^{-1} D x_t. \tag{6.6}$$

The matrix decomposition mentioned in (6.5) shows that the sign-solvability conditions of the model in (6.3) can also be analysed in two steps, viz. for vector y_1 and y_2 successively. The first step in the sign-solvability approach to solve for vector y_1 is based on:

$$A_{11} y_{1,t} = B_{11} y_{1,t-1} + C_{11} y_{1,t-2} + d_{11} RA_t + d_{12} P_t, \tag{6.7}$$

or, because matrix A_{11} is non-singular:

$$y_{1,t} = A_{11}^{-1} B_{11} y_{1,t-1} + A_{11}^{-1} C_{11} y_{1,t-2} + A_{11}^{-1} d_{11} RA_t + A_{11}^{-1} d_{12} P_t, \tag{6.8}$$

with y_{1t} a vector incorporating the variables SA, EA, REC and EMP for pe-

riod t. Sign-solvability of the vector y_{1t} may exist, independently of the
fact whether the conditions of sign-solvability hold for the vector y_{2t}.
The vector y_{2t} is a vector incorporating the variables CR, CO, ER and EO.
This vector of variables refers to the second part of the equations in (6.5),
viz.:

$$A_{21} y_{1,t} + A_{22} y_{2,t} = B_{21} y_{1,t-1} + B_{22} y_{2,t-1} + d_{21} RA_t + d_{22} P_t, \qquad (6.9)$$

or, since matrix A_{22} is non-singular:

$$y_{2,t} = -A_{22}^{-1}A_{21}y_{1,t} + A_{22}^{-1}B_{21}y_{1,t-1} + A_{22}^{-1}B_{22}y_{2,t-1} + A_{22}^{-1}d_{21}RA_t + A_{22}^{-1}d_{22}P_t \qquad (6.10)$$

A graph representation concerning the qualitative impacts between the variables SA, EA, REC, EMP, CR, CO, ER and EO - denoted by matrix A in equation
(6.4) - is depicted in Figure 6.6.

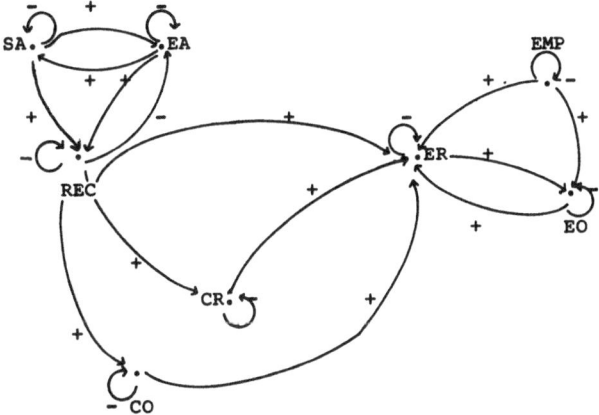

Figure 6.6. Graph representation of the qualitative impacts between the
variables of matrix A.

The qualitative impacts in Figure 6.6 show whether the inverse of matrix A
from equation (6.4) can be determined uniquely when only qualitative information is available. Matrix A can be inverted in a qualitative way when the
two first conditions of sign-solvability (discussed in subsection 5.2.3) hold
in the graph representation depicted in Figure 6.6.
The first condition of sign-solvability holds for all variables in Figure
6.6, because all main diagonal elements of the matrix A are negative. However, the second condition of sign-solvability, viz. that all cycles of at

least length two be non-positive, does not hold for the graph representation in Figure 6.6. This can easily be seen because the cycles SA-EA-SA and ER-EO-ER, which are cycles with length two, are both positive.

The inverse of the matrix A is therefore not defined up to the signs of cell-elements and the model in (6.4) is not sign-solvable, i.e., the vector y_t cannot be expressed in qualitative terms by means of y_{t-1}, y_{t-2} and x_t just like the reduced-form model in (6.6).

Figure 6.6 also shows that partial sign-solvability with qualitative information does not exist either, because the second condition of sign-solvability does not hold for the variables SA, EA, and REC.

However, a systems model like the IEM systems model as mentioned in (6.3), which is not sign-solvable in a pure qualitative sense, may become at least partially sign-solvable, when a mixture of qualitative and quantitative information is available. Such quantitative information concerning parameter values may be obtained from prior knowledge or theoretical evidence, or be based on estimates or assessments.

Parameter assessments of the simulation model (6.3) are presented in Table 6.2.

Table 6.2. Parameter assessment of the simulation model.

α	β	γ	δ	ε	η	λ	μ
0.01	0.5	0.001	0.025	0.05	0.05	0.05	0.01
0.001	0.05	0.9			0.04	0.02	0.01
0.1	1.05	0.4				0.1	
	0.2	0.1				0.1	
		0.05				0.05	
						0.01	

The sign-values of the reduced-form equations in model (6.6) are evidently uniquely determined when the parameter assessments from Table 6.2 are included. The sign-values of the reduced form in equation (6.8) then become:

$$
\begin{bmatrix} SA \\ EA \\ REC \\ EMP \end{bmatrix}_t =
\begin{bmatrix} + & + & - & 0 \\ + & + & - & 0 \\ + & - & + & 0 \\ 0 & 0 & 0 & + \end{bmatrix}
\begin{bmatrix} SA \\ EA \\ REC \\ EMP \end{bmatrix}_{t-1} +
\begin{bmatrix} 0 & - & + & 0 \\ 0 & - & + & 0 \\ 0 & + & + & 0 \\ 0 & 0 & 0 & 0 \end{bmatrix}
\begin{bmatrix} SA \\ EA \\ REC \\ EMP \end{bmatrix}_{t-2} +
\begin{bmatrix} + & - \\ + & - \\ + & + \\ 0 & + \end{bmatrix}
\begin{bmatrix} RA_t \\ P_t \end{bmatrix}, \quad (6.11)
$$

or:

$$y_{1,t} = M_1 y_{1,t-1} + M_2 y_{1,t-2} + M_3 x_t \qquad (6.12)$$

The sign-values of the reduced form in equation (6.10) then become:

$$
\begin{bmatrix} CR \\ CO \\ ER \\ EO \end{bmatrix}_t =
\begin{bmatrix} + & 0 & 0 & 0 \\ 0 & + & 0 & 0 \\ 0 & 0 & + & 0 \\ 0 & 0 & 0 & + \end{bmatrix}
\begin{bmatrix} CR \\ CO \\ ER \\ EO \end{bmatrix}_{t-1} +
\begin{bmatrix} + & - & - & 0 \\ + & - & - & 0 \\ + & - & - & 0 \\ + & - & - & 0 \end{bmatrix}
\begin{bmatrix} SA \\ EA \\ REC \\ EMP \end{bmatrix}_{t-1} +
\begin{bmatrix} 0 & + & + & 0 \\ 0 & + & + & 0 \\ 0 & + & + & 0 \\ 0 & + & + & 0 \end{bmatrix}
\begin{bmatrix} SA \\ EA \\ REC \\ EMP \end{bmatrix}_{t-2} +
\begin{bmatrix} + & + \\ + & + \\ + & + \\ + & + \end{bmatrix}
\begin{bmatrix} RA_t \\ P_t \end{bmatrix}
$$

or:

$$y_{2,t} = K_1 y_{2,t-1} + K_2 y_{1,t-1} + K_3 y_{1,t-2} + K_4 x_t. \qquad (6.14)$$

The reduced form equations in (6.11) and (6.13) - denoted in qualitative terms - for example - show, that an increase of environmental attractiveness (EA) during period t-1 will in the next period lead to an increase of the following variables: stimulating efforts for regional attractiveness (SA) and environmental attractiveness (EA); it will also lead to a decrease in value of the following variables: demand for recreational facilities (REC), expenditure for consumption goods (CR) and non-consumption goods (CO) by recreationers as well as the demand for employment, either related to recreational activities (ER) or to 'other' activities (EO).

An increase of population size (P) during period t will lead to a decrease of the variables SA and EA, and to an increase in value of all other endogenous variables of the IEM which were specified in model (6.1).

However, some mathematical tools discussed in subsection 5.2.3 may become helpful in the sign-solvability process to rewrite model (6.2) into (6.11) and (6.13). The first one is a matrix permutation and matrix decomposition procedure to rearrange equations and variables to give a hierarchical model representation (see equations (5.7) and (5.8) in subsection 5.2.3). The simulation model in (6.2) is already written in a hierarchical form (see for example Table 6.1 with a hierarchical representation of the variables and lower level variables which are an input to higher level variables). Sign-solvability of the vector y1 with variables SA, EA, REC and EMP in equation

(6.11) can be analysed independently of the other variables. However, the graph representation of the qualitative impacts between variables, depicted in Figure 6.6, shows that model (6.7) is not sign-solvable with purely qualitative information, because the second condition of sign- solvability does not hold either.

The second point mentioned in subsection 5.2.3 deals with the use of a mixture of qualitative and quantitative information, and makes a distinction in a top-down and a bottom-up approach, so as to assure that a qualitative systems model may become sign-solvable in a number of steps by means of a sequential introduction of numerical information on parameter values.

The inverse of the matrix A_{11} in equation (6.7) has to be determined in the reduced-form representation in equation (6.8), and is equal to:

$$A_{11}^{-1} = \frac{1}{\det(A_{11})} \begin{bmatrix} -(1+\beta_2\gamma_2) & -\alpha_1 & \alpha_1\beta_1 & 0 \\ \beta_2\gamma_2-\beta_1 & -1 & \beta_2 & 0 \\ -(\gamma_1+\beta_1\gamma_2) & -(\gamma_2+\alpha_1\gamma_1) & -(1-\alpha_1\beta_1) & 0 \\ 0 & 0 & 0 & -\det(A_{11}) \end{bmatrix}, \quad (6.15)$$

where $\det(A_{11})$ is the determinant value of the matrix A_{11}, which is equal to:

$$\det(A_{11}) = 1 + \beta_2\gamma_2 + \beta_2\gamma_1\alpha_1 - \alpha_1\beta_1. \quad (6.16)$$

The inverse of matrix A_{11} in equation (6.15) shows that quantitative information concerning the parameter values of the equations for the endogenous variables SA, EA and REC is necessary in order to determine whether the variables from the reduced form equations can be solved in a qualitative way.

Eight steps for the sign-solvability analysis of the simulation model are presented in Table 6.3 below by means of a sequential introduction of quantitative information. This stepwise procedure shows the variables which are relevant with respect to the sign-solvability analysis of the simulation model. The undetermined cell-entries of the matrices M_1, M_2, M_3, K_1, K_2, K_3, and K_4 (defined in respectively equations (6.12) and (6.14)) are computed in each step when the parameter assessments from various parameters are used. An empty block in any column of Table 6.3 means that the corresponding matrix has no cell-entries which are undetermined regarding their signs. The use of quantitative information when moving from step 1 to step 8 in Table 6.3 shows that an originally not-sign-solvable system with quantitative information regarding the parameters α, β and γ (in step 1) may become sign- solvable when additional quantitative information regarding the parameters λ and μ (in step 8) is also included.

The table shows that the vector y_1 in equation (6.12) with variables SA, EA, REC, and EMP is sign-solvable when quantitative information of the parameters α, β, and γ is used (see step 1 in the table). It also shows that the use of more quantitative information will not necessarily improve sign-solvability. Consider for example the results of step 1 and step 2, with the difference that the parameter assessment of $\delta_1 = 0.025$ is introduced in step 2. However, both analyses have the same unknown cell-entries from matrices K_2, K_3 and K_4. The same conclusion holds by comparing step 1 and step 3 (when the parameter assessment of ε_1 is included), step 1 and step 4 (by inclusion of the para- meter assessment of η_1 and η_2), step 1 and step 5 (by inclusion of ε_1, η_1 and η_2), step 1 and step 6 (by inclusion of λ_1, λ_2, λ_3, λ_4, λ_5, and λ_6) and step 1 and step 7 (by inclusion of μ_1 and μ_2). It appears from the analysis that sign-solvability exists for the simulation model when quantitative informa- tion is used with respect to the parameters mentioned in step 8 from Table 6.3. This means that information on the signs of the parameters ε_1, δ_1, η_1, and η_2 (all considered to be positive) is sufficient for the sign-solvability procedure.

The use of a matrix decomposition or matrix permutation procedure as well as of available quantitative information provides new tools for a qualitative policy impact analysis. A stepwise procedure - such as the one presented in Table 6.3 - may be employed in order to obtain solutions for the sign-solva- bility approach.

Qualitative calculus, and especially sign-solvability analysis, may solve IEMs with a dynamic nature and qualitative information regarding the impacts between variables. The tools discussed in this section in the frame of a dynamic simulation model are relevant in modelling environmental phenomena to achieve sign-solvability.

Table 6.3. Mixed level of information and sign-solvability analysis.

		Unknown signs of cell-entries						
Step	Information	M_1	M_2	M_3	K_1	K_2	K_3	K_4
1	$\alpha_1,\alpha_2,\alpha_3$ $\beta_1,\beta_2,\beta_3,\beta_4$ $\gamma_1,\gamma_2,\gamma_3,\gamma_4$ γ_5					(3,1),(3,2),(3,3) (4,1),(4,2),(4,3)	(3,2),(3,3) (4,2),(4,3)	(3,1),(3,2) (4,1),(4,2)
2	$\alpha_1,\alpha_2,\alpha_3$ $\beta_1,\beta_2,\beta_3,\beta_4$ $\gamma_1,\gamma_2,\gamma_3,\gamma_4$ γ_5,δ_1					(3,1),(3,2),(3,3) (4,1),(4,2),(4,3)	(3,2),(3,3) (4,2),(4,3)	(3,1),(3,2) (4,1),(4,2)
3	$\alpha_1,\alpha_2,\alpha_3$ $\beta_1,\beta_2,\beta_3,\beta_4$ $\gamma_1,\gamma_2,\gamma_3,\gamma_4$ γ_5,ε_1					(3,1),(3,2),(3,3) (4,1),(4,2),(4,3)	(3,2),(3,3) (4,2),(4,3)	(3,1),(3,2) (4,1),(4,2)
4	$\alpha_1,\alpha_2,\alpha_3$ $\beta_1,\beta_2,\beta_3,\beta_4$ $\gamma_1,\gamma_2,\gamma_3,\gamma_4$ γ_5,η_1,η_2					(3,1),(3,2),(3,3) (4,1),(4,2),(4,3)	(3,2),(3,3) (4,2),(4,3)	(3,1),(3,2) (4,1),(4,2)
5	$\alpha_1,\alpha_2,\alpha_3$ $\beta_1,\beta_2,\beta_3,\beta_4$ $\gamma_1,\gamma_2,\gamma_3,\gamma_4$ $\gamma_5,\varepsilon_1,\eta_1,\eta_2$					(3,1),(3,2),(3,3) (4,1),(4,2),(4,3)	(3,2),(3,3) (4,2),(4,3)	(3,1),(3,2) (4,1),(4,2)
6	$\alpha_1,\alpha_2,\alpha_3$ $\beta_1,\beta_2,\beta_3,\beta_4$ $\gamma_1,\gamma_2,\gamma_3,\gamma_4$ γ_5 $\lambda_1\ \lambda_2,\lambda_3,\lambda_4$ λ_5,λ_6					(3,1),(3,2),(3,3) (4,1),(4,2),(4,3)	(3,2),(3,3) (4,2),(4,3)	(3,1),(3,2) (4,1),(4,2)
7	$\alpha_1,\alpha_2,\alpha_3$ $\beta_1,\beta_2,\beta_3,\beta_4$ $\gamma_1,\gamma_2,\gamma_3,\gamma_4$ γ_5,μ_1,μ_2					(3,1),(3,2),(3,3) (4,1),(4,2),(4,3)	(3,2),(3,3) (4,2),(4,3)	(3,1),(3,2) (4,1),(4,2)
8	$\alpha_1,\alpha_2,\alpha_3$ $\beta_1,\beta_2,\beta_3,\beta_4$ $\gamma_1,\gamma_2,\gamma_3,\gamma_4$ γ_5 $\lambda_1,\lambda_2,\lambda_3,\lambda_4$ $\lambda_5,\lambda_6,\mu_1,\mu_2$							

6.5. CONCLUDING REMARKS

The design of an integrated modelling approach was developed in this chapter
with a modular representation of the phenomena concerned. The phenomena selected in case of the Biesbosch area are basically directly or indirectly
related to the recreational activities in that area. The phenomena which are
considered to be relevant for the recreational level (pattern and behaviour)
are (regional) economic activities, demographic and environmental phenomena.
The first topic mentioned in this chapter was the presentation of the outline
of such an IEM in terms of the satellite design of integration, and which was
actually showed to be a hierarchical systems representation with lower level
links between modules also being covered.
The links between variables are interpreted in terms of the model interdependencies, and such interdependencies will trace out a hierarchical structure
if that in fact exists. The structure of the links mentioned are being analysed in a systematic way for the graph approach. A simulation model is developed which is based on the links between variables from the systems design. A
sign-solvability approach has been presented to determine the qualitative
impacts of the lagged endogenous variables as well as the exogenous variables
on the endogenous variables. However, this sign-solvability approach becomes
especially relevant for integrated environmental modelling when the hierarchical outline of the stimulus-response relationships will be used to trace
partial sign-solvability, and when a mixture of qualitative and quantitative
information is available, because of the severe restrictions on sign-solvability when only qualitative information is available.
The graph approach for the analysis of the binary relationships has shown its
relevance in this chapter for the investigation of an hierarchical model
structure which will be used in the sign-solvability approach to determine
the qualitative model representation.
Having discussed in this chapter the design and the stimulus-response relationships of an IEM approach with various levels of information, some tools
will be further elaborated in the next chapter with respect to the operationalization of the modules from Figure 6.3.

CHAPTER 7. OUTDOOR RECREATION IN THE BIESBOSCH AREA

7.1. INTRODUCTION

Outdoor recreational activities are connected with the supply of and the demand for natural resources. Examples of outdoor recreation which may stress natural environment are visits to national parks, and activities in lakes such as fishing, swimming or canoeing (Murphy, 1985). Natural resources which may be relevant for recreational purposes are, among others, forests, canyons, lakes or bays (see also McConnell, 1985 for an exposition of outdoor recreation in the frame of natural resources). The definition of outdoor recreation with its dependency on natural resources means that recreational activities such as tennis or golf are of only limited relevance, because usually they do not have direct links to natural resources.

The regional-economic relationship between outdoor recreation and nature conservation will be analysed in this chapter by means of a model framework for the Biesbosch region in the Netherlands, in terms of the perception, the preference, and the motivation of recreationers to spend their leisure time in that area.

Recreational activities are the main elements of the analysis, in accordance with the outline of the satellite model design of integration, as it has been presented in Figure 6.3 at the modular level and in Figure 6.4 at the level of variables. A survey analysis is a useful tool to explore the perception, the preference, and the behaviour of recreationers in a spatial context with respect to outdoor recreation, economic activities and nature conservation. Therefore a survey was organized in the Biesbosch area during the summer of 1983 concerning the spatial pattern of the recreational activities. The total sample size of the survey was about 400 respondents. The Biesbosch area is subdivided into 5 regions to be able to present a spatial characterization of outdoor recreation. A counting was also organised in that period with respect to type of boat, use of banks and type of recreational activities. The Biesbosch area is subdivided into 11 regions.

The satellite design of integration of an IEM for the Biesbosch area was presented in Figure 6.3 in terms of modules. The recreation module is the core component and three other modules (viz., demography, regional economy, and natural environment) depict the lower levels in this hierarchical model outline, because they are based upon recreational activities.

Some graph-theory based tools have been used in the previous chapter, to elaborate an IEM when binary or qualitative terms are the only available information with respect to the impacts between variables. The tools used in that chapter are an analysis of the causal model structure, reflecting the nature of model dependencies and interdependencies, and an analysis of a model represented in qualitative terms. These tools are especially relevant in the subject of modelling environmental phenomena, due to the scarcity of reliable information and adequate experiments concerning the time pattern of the variables.

However, some additional topics were presented in Sections 5.3 and 5.4 upon the integration of economic and environmental phenomena, viz. the non-metric or discrete nature of information from surveys, and the multivariate nature of information in multidisciplinary systems.

The aim of this chapter is to analyse the outdoor recreation in the Biesbosch area in a spatial context. The level of recreational activities may be determined by various environmental phenomena such as, for example, the characteristics of landscape or the presence of plants and animals, and by the presence of recreational facilities. The various tools which have been presented in Sections 5.3 and 5.4 are being used in the next sections to analyse, among others, the demographic nature of the recreationers, and the relevance of environmental characteristics for recreational activities.

The first issue, to be discussed in Section 7.2, is an exploratory analysis on the demographic and socio-economic characteristics of the recreationers. The spatial scale covers the various subregions of the Biesbosch area. Outdoor recreation also has an economic component, and the place where the recreationers from the survey make their daily purchases will be analysed also. The place where the recreationers make their daily purchases will be cross-classified with the duration of their stay and the distance between the home address and the Biesbosch area.

The second issue, to be discussed in Section 7.3, has to do with the relevance of various phenomena to spend a period in the Biesbosch area. The phenomena which may determine the choice of residence are, among others, the availability of recreational facilities, the nature of landscape, the ecological characteristics or the distance to the home address. The phenomena mentioned are represented in a spatial context, and be analysed by means of the HOMALS program.

The specific natural resources in the Biesbosch area, with respect to outdoor recreational activities, are the type of banks where boats can be laid (willow-trees, reed vegetation, landing-stages, beaches, e.g.). The banks are being used by different types of boats (either motorized or unmotorized boats), and they may also be used by the recreationers for a wide range of activities (among others, fishing, swimming, walking, or playing football). The spatial phenomena are therefore multivariate in nature, and a multidimensional scaling procedure is presented also in this section with respect to these variables. The purpose is to define the underlying structure from the multivariate phenomena. The information is obtained from counts at a spatial disaggregated scale with the Biesbosch area being subdivided into 11 sub-regions.

The causal structure between the type of boat, the type of banks where the boats can be laid and the type of recreational activities, will be analysed in Section 7.4 by estimating a path model.

The relevance of the various tools, which are presented in this chapter within the framework of an integrated environmental modelling approach with respect to outdoor recreation in the Biesbosch area, will be evaluated in Section 7.5.

7.2. A SPATIAL CHARACTERIZATION OF OUTDOOR RECREATION

The demographic nature and the perception of recreationers concerning outdoor recreation in the Biesbosch area will be explored in this section in a spatial context. As already mentioned in the introductory section, a survey was organized during the summer of 1983 with recreationers who spent their leisure time in the Biesbosch area (Van der Linden and Van Eijk, 1984). The survey was held during six days to characterize the various periods in the summer (i.e., the days before and during public holidays), and also in five sub-areas to characterize the various areas in the Biesbosch.

A characterization of the respondents from the survey according to the areas of visit and the period of the year is presented in Table 7.1.

Table 7.1. Respondents in the survey according to the area of visit and the period of the year.

Area Day	June 11	June 25	July 13	July 16	August 10	August 13	Total
Rietplaat	35	19	15	14	35	0	118
Honderddertig	1	10	3	1	2	2	19
Keesjes Killeke	40	23	25	28	26	0	142
Merwelanden	16	24	4	28	6	0	78
Zuidhollandse Biesbosch	8	12	0	1	5	2	28
Total	100	88	47	72	74	4	385

The Rietplaat area is located in the central part of the Biesbosch region, having a lot of recreational facilities with beaches and picknick places, as well as wide water with good sailing facilities. The area named Honderddertig

is very quiet in nature, because it has only limited entrance possibilities
for motorized boats. The quiet nature of this area may be a well founded
reason of the small number of respondents in the survey of only 19 out of the
total sample size of 385. The area Keesjes Killeke is located in the centre
of the Biesbosch region with all types of boats allowed to stay there. However, this area differs from the Rietplaat area because it has no recreational facilities.

The areas Merwelanden and Zuidhollandse Biesbosch have been chosen to be
included in the survey, since these two areas are spatially separated from
the three other areas. These two areas may therefore show a difference in
recreational patterns. The accessibility of the Zuidhollandse Biesbosch, for
example, is rather poor, and the recreational activities in this area may for
that reason be rather small in nature.

Figure 7.1 shows the percentage distribution of the distance between the home
address and the Biesbosch area of the respondents. The percentage distribution is presented for each of the five regions of the survey, as well as for
the total sample size.

Figure 7.1 shows that the major part of the recreationers live at a distance
of less than 20 kilometers from the Biesbosch area. For the area Merwelanden
even more than 50 percent of the respondents stay less than 10 kilometers
from their home address. This phenomenon indicates that the Biesbosch area
especially has a regional relevance for recreationers.

A classification of the recreationers of the survey with respect to their age
and their level of education is presented in Table 7.2 in a multidimensional
contingency table with size 5x3x3. The variables 'area of visit', 'age' and

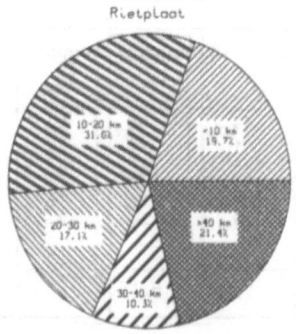

Fig. 7.1a. Distance between Rietplaat
and home address.

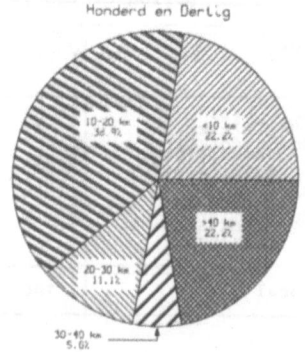

Fig. 7.1b. Distance between
Honderddertig and
home address.

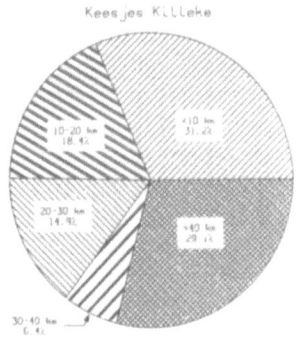

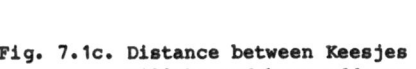

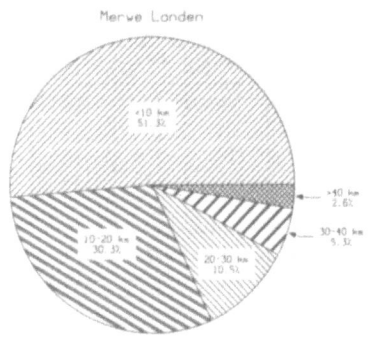

Fig. 7.1c. Distance between Keesjes
            Killeke and home address.

Fig. 7.1d. Distance between Merwe-
            landen and home address.

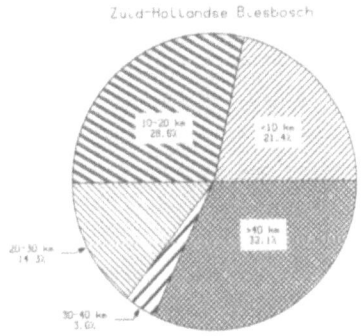

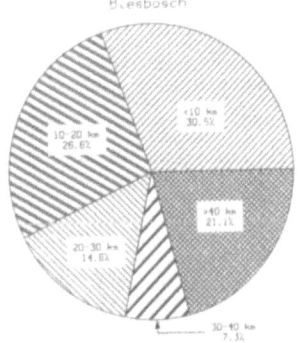

Fig. 7.1e. Distance between Zuidhollandse Fig. 7.1f. Distance between Bies-
            Biesbosch and home address.   bosch and home address.

Figure 7.1. Distance between Biesbosch area and home address for recreationers.

'level of education' are successively represented by means of A, B, and C.
An exploratory analysis of the interdependence between the variables from
Table 7.2 can be expressed adequately in terms of log-linear models with
main effects and interaction effects. The cell-frequencies which are zero are
considered to be zero because of the small sample size, and not because of

Table 7.2. A multidimensional table of recreationers, cross-classified by
their age and their level of education.

	C = 1			C = 2			C = 3			Total
	B=1	B=2	B=3	B=1	B=2	B=3	B=1	B=2	B=3	
A=1	9	29	18	10	23	3	4	12	1	109
A=2	0	3	1	4	1	0	1	7	0	17
A=3	10	26	19	17	21	12	13	13	6	137
A=4	7	15	28	5	8	5	3	2	3	76
A=5	2	4	5	4	2	4	1	3	1	26
Total	28	77	71	40	55	24	22	37	11	365

Missing values: 24

A = Area: 1 = Rietplaat; B = Age: 1 = < 29 years;
          2 = Honderddertig; 2 = 30-50 years;
          3 = Keesjes Killeke; 3 = > 50 years.
          4 = Merwelanden;
          5 = Zuidhollandse Biesbosch. C = Education: 1 = primary school;
                                        2 = secondary school;
                                        3 = higher education.

so-called systematic zeroes. The use of log-linear models is an appropriate
way in case of spatial information to test hypotheses about interactions
among variables (see also subsection 5.3.3 for an hierarchical set of log-
linear models with multidimensional contingency tables and nominal information, and Whittam and Siegel-Causey, 1981 for some examples to analyse species interactions and community structure by making use of log- linear models).

Table 7.3 shows the goodness-of-fit statistics from an hierarchical set of
log-linear models, which have been fitted to the information of Table 7.2.
The models are estimated by means of the ECTA computer package with an iterative proportional fitting procedure. The critical values shown in the last
column of Table 7.3 are based upon the degrees of freedom which remain after
having included the mentioned parameters, and the 95 percent level of significance.
All 18 possible models in the hierarchical set of log-linear models, with
respect to a three-dimensional table, are presented in Table 7.3.
Two model selection procedures will be used now to find the most parsimonious
member from the set of all possible hierarchical log-linear models in a systematic way. The selection strategies - to select the models which have a
satisfactory fit to the observed data - are successively screening and a
stepwise selection procedure (see also the presentation of the selection
procedures in subsection 5.3.4).

Table 7.3. Members of the set of hierarchical log-linear models fitted to
          Table 7.2.

Model	G^2	degrees of freedom	critical value
A + B + C + A×B + A×C + B×C	17.33	16	26.30
A + B + C + A×C + B×C	39.59	24	36.42
A + B + C + A×B + B×C	34.65	24	36.42
A + B + C + A×B + A×C	39.34	20	31.41
A + B + C + B×C	62.35	32	46.17
A + B + C + A×C	67.04	28	41.34
A + B + C + A×B	62.11	28	41.34
B + C + B×C	225.13	36	50.97
A + C + A×C	94.62	30	43.77
A + B + A×B	109.40	30	43.77
A + B + C	89.81	36	50.97
B + C	252.58	40	43.77
A + C	117.39	38	53.36
A + B	137.10	38	53.36
C	280.16	42	58.11
B	299.87	42	58.11
A	164.68	40	55.76
Equiprobability	327.46	44	60.44

Screening is a selection procedure which makes use of the partial and the
marginal association test statistics, and these two tests can be thought of
as lower and upper bounds of the conditional G^2-values. The partial and marginal association values for each of the terms, their degrees of freedom, and
the critical values at the 95 percent level of significance are presented in
Table 7.4.

Table 7.4. Partial and marginal association test statistics.

λ-term	d.f.	Partial association	Marginal association	Critical value
A	4	162.78	162.78	9.49
B	2	27.59	27.59	5.99
C	2	47.30	47.30	5.99
A×B	8	12.26	27.70	15.5
A×C	8	17.32	22.77	15.5
B×C	4	12.01	27.45	9.49
A×B×C	16	17.33	17.33	26.3

When the partial and the marginal test statistics are both large, the corresponding term then would appear to be required in the model. The main effect
terms A, B and C, and the first-order interaction terms A×C and B×C are
therefore likely to be included in the selected model. The first-order interaction term A×B is in need of further investigation, because its partial

association is smaller than the critical value, and its marginal association is larger than the critical value. The second-order interaction term AxBxC is unlikely to be included because both of its test-statistics are smaller than the critical value.

The model A+B+C+AxC+BxC will be chosen as the base model in the screening procedure, and the possibility of adding the first-order interaction term between the variables A and B will be investigated with a forward selection procedure. The inclusion of the AxB interaction term will lead to a signifi-cant decrease in the G^2-value from the base model. The model selected there-fore by screening is A+B+C+AxB+AxC+BxC with a G^2-value of 17.33, which is a model with 16 degrees of freedom and within the range of acceptance.

The second model selection procedure is termed <u>stepwise selection</u> of log-linear models, and it includes both the forward selection and the backward elimination procedures as was already mentioned in subsection 5.3.4.

The forward selection procedure starts in this case with the log-linear model containing all main effects as the base model, and this procedure selects the model A+B+C+AxB+BxC. The backward elimination procedure starts with the log-linear model containing all first-order interaction effects as the base mo-del. The model selected with this selection procedure which fits the data well is also A+B+C+AxB+BxC.

Table 7.5. Estimates of the main effects and first-order interaction effects (with 'centred-effect' constraints).

Parameter	Magnitude of the effect	Standardised value	Parameter	Magnitude of the effect	Standardised value
λ^A (1)	.679	4.551	λ^{AB} (1,1)	-.161	-.823
λ^A (2)	-1.447	-3.752	λ^{AB} (2,1)	.439	.987
λ^A (3)	1.015	7.385	λ^{AB} (3,1)	.056	.319
λ^A (4)	.387	2.519	λ^{AB} (4,1)	-.297	-1.406
λ^A (5)	-.634	-3.076	λ^{AB} (5,1)	-.037	-.132
λ^B (1)	-.076	-.543	λ^{AB} (1,2)	.267	1.547
λ^B (2)	.507	3.908	λ^{AB} (2,2)	.632	1.520
λ^B (3)	-.431	-2.141	λ^{AB} (3,2)	-.134	-.824
λ^C (1)	.434	3.178	λ^{AB} (4,2)	-.382	-2.002
λ^C (2)	.079	.515	λ^{AB} (5,2)	-.382	-1.435
λ^C (3)	-.513	-2.752	λ^{AB} (1,3)	-.106	-.412
λ^{BC} (1,1)	-.472	-2.631	λ^{AB} (2,3)	-1.071	-1.482
λ^{BC} (2,1)	-.077	-.486	λ^{AB} (3,3)	.078	.331
λ^{BC} (3,1)	.549	2.359	λ^{AB} (4,3)	.679	2.751
λ^{BC} (1,2)	.239	1.278	λ^{AB} (5,3)	.419	1.296
λ^{BC} (2,2)	-.059	-.333			
λ^{BC} (3,2)	-.180	-.660			
λ^{BC} (1,3)	.233	1.040			
λ^{BC} (2,3)	.136	.647			
λ^{BC} (3,3)	-.369	-1.094			

The most parsimonious model which fits the information from Table 7.2, is achieved by the stepwise selection procedure. Parameter estimates for the main effects and the first-order interaction effects from that model, as well as their standardised values, are summarised in Table 7.5.

As already mentioned in equation (5.22), the main effect parameters and the first-order interaction effect parameters are defined as deviations from the overall-mean effect λ, which is equal to 1.564.

A spatial characterization for each of the five Biesbosch sub-areas concerning the age and the educational level of the recreationers can be obtained from the parameter estimates mentioned in Table 7.5. This will be summarised in Table 7.6.

Table 7.6. Spatial characterization of the recreationers out of the survey, with respect to their age and their level of education.

Area	Characterization of the interaction effect AxB
Rietplaat	The number of recreationers in the age group between 30 and 50 years is above the average pattern of the survey.
Honderddertig	The number of recreationers in the age group below 30 years and between 30 and 50 years is in this area far above the average pattern of the survey.
Keesjes Killeke	The age of the recreationers for the three groups is around the average, with the exception of the age group between 30 and 50 years which is in this area only slightly below the average.
Merwelanden	The number of recreationers in the age group of more than 50 years is above the average pattern in this area.
Zuidhollandse Biesbosch	Same conclusion concerning the age of recreationers visiting the area Merwelanden, with the exception that the age group of recreationers until 30 years is now just around the average of the survey.

Characterization of the dependency between age and educational level (the interaction effect BxC).

Age group of recreationers less than 30 years: the recreationers with at least secondary school are above the average pattern in this age group.

Age group of recreationers between 30 and 50 years: the recreationers with higher education are above the average in this group, and the recreationers with lower educational levels are slightly below the average in this age group.

Age group of recreationers above 50 years: the respondents with only primary school are far above the average level in this age group, and the opposite direction may be indicated from the survey for the two other educational levels.

A four-way contingency table of size 5×2×2×2 concerning the variables 'area of visit', 'type of boat', 'place to spend the night' and 'distance between home address and Biesbosch area' is depicted in Table 7.7, with variables represented by A, B, C, and D respectively.

The interdependencies between the variables from Table 7.7 will be expressed in terms of an exploratory analysis of log-linear models with main effects and interaction effects. The total number of possible models in hierarchical log-linear models in case of a four- dimensional contingency table is 167 (see Wrigley, 1985, p. 174, where 28 model types of the hierarchical set of log- linear models are depicted in case of a four-dimensional contingency table). In such case, a systematic and efficient model selection strategy is required.

Commonly used criteria are parsimony of parameters, goodness-of-fit, simplicity of interpretation, and the inclusion of the significant effects.

The model selection procedures of the log-linear models to fit the data from Table 7.7 are respectively screening and a stepwise selection procedure.

Table 7.7. A multidimensional table of recreationers, cross-classified by type of boats, where to spend the night and the distance to home address.

	D = 1				D = 2				Total
	C = 1		C = 2		C = 1		C = 2		
	B=1	B=2	B=1	B=2	B=1	B=2	B=1	B=2	
A=1	8	4	62	6	4	1	28	3	116
A=2	0	4	6	5	0	1	2	2	20
A=3	8	20	58	3	5	5	27	13	139
A=4	10	10	43	6	0	0	5	1	75
A=5	2	2	10	4	1	3	3	3	28
Total	28	40	179	24	10	10	65	22	378

Missing values: 21

A = Area: 1 = Rietplaat;   B = Type of boat: 1=speed-boat or motor-boat;
         2 = Honderddertig;   2=sail-boat, row-boat or canoe.
         3 = Keesjes Killeke;
         4 = Merwelanden;
         5 = Zuidhollandse Biesbosch.
C = Place to spend the night: 1 = spend the night at home;
                               2 = spend the night in the Biesbosch area.
D = Distance between home address and Biesbosch area: 1 = < 30 kilometers;
                                                        2 = > 30 kilometers.

The partial and marginal association test statistics, which are relevant in case of a screening selection procedure, to determine the significance of each effect, are presented in Table 7.8.

Table 7.8. Partial and marginal association test statistics.

λ-term	d.f.	Partial association	Marginal association	Critical value
A	4	158.63	158.63	9.49
B	1	95.63	95.63	3.84
C	1	113.78	113.78	3.84
D	1	73.58	73.58	3.84
A×B	4	25.10	28.88	9.49
A×C	4	3.84	7.51	9.49
A×D	4	23.92	24.34	9.49
B×C	1	53.00	54.36	3.84
B×D	1	3.45	1.56	3.84
C×D	1	3.82	1.81	3.84
A×B×C	4	6.17	3.68	9.49
A×B×D	4	4.31	2.22	9.49
A×C×D	4	7.01	5.41	9.49
B×C×D	1	6.94	4.41	3.84
A×B×C×D	4	3.39	3.39	9.49

Each term in Table 7.8 can be screened from the partial and the marginal association test statistics and belongs to one of the following three classes:

(i)   likely to be important and required to be included in the model. This holds for the main-effect terms A, B, C and D, the first-order interaction effects A×B, A×D and B×C, and the second-order interaction effect B×C×D;

(ii)   likely to be unimportant and not required to be included in the model, which holds for the first-order interaction effects A×C, B×D and C×D, the second-order interaction effects A×B×C, A×B×D, A×C×D, and the third-order interaction effect A×B×C×D;

(iii) in need of further investigation, which does not hold for any interaction term from Table 7.8.

The base model in this screening procedure includes the terms of the hierarchical log-linear model, which are likely to be important and require to be included in the model, and is:

$$\log_e m_{ijkl} = \lambda + \lambda_i^A + \lambda_j^B + \lambda_k^C + \lambda_l^D + \lambda_{ij}^{AB} + \lambda_{il}^{AD} + \lambda_{jk}^{BC} + \lambda_{jl}^{BD} + \lambda_{kl}^{CD} + \lambda_{jkl}^{BCD}, \quad \begin{array}{l} i=1,\ldots,5 \\ j,k,l = 1,2 \end{array} \qquad (7.1)$$

with G^2-value equal to 22.25 and 20 degrees of freedom. The χ^2-critical value at the 95 percent level of significance and 20 degrees of freedom is 31.41, and the base model is therefore selected from the screening procedure as a model which fits the data well.

The stepwise selection procedure has the model with all main effects as the base model. The goodness-of-fit statistics from the forward selection procedure, which include those first-order interaction effects to the base model giving a significant decrease in G^2-values, are presented in Table 7.9.

The first-order interaction effects, which are added to the base model with all main effects in the forward selection procedure are successively BxC, AxB and AxD. The model chosen from this selection procedure which fits the data from Table 7.7, now becomes:

$$\log_e m_{ijkl} = \lambda + \lambda_i^A + \lambda_j^B + \lambda_k^C + \lambda_l^D + \lambda_{ij}^{AB} + \lambda_{il}^{AD} + \lambda_{jk}^{BC} , \quad \begin{array}{l} i=1,\ldots,5 \\ j,k,l = 1,2 \end{array} \quad (7.2)$$

The information of Table 7.7 has been fitted now by making use of the iterative proportional fitting procedure, and two model selection procedures (viz. screening and the forward selection procedure). The most parsimonious model which fits that data well will be the one in equation (7.2), and the interdependency between the variables 'type of boat', 'place to spend the night', as well as 'the distance to the home address of the recreationers'

Table 7.9. Stepwise forward selection procedure.

Model specification	G^2	d.f.	Conditional G^2	d.f.
(1): A+B+C+D	139.73	32		
(1)+AxB	110.84	28	28.89	4
(1)+AxC	132.21	28	7.52	4
(1)+AxD	115.39	28	24.34	4
(1)+BxC	85.37	31	54.36	1
(1)+BxD	138.16	31	1.57	1
(1)+CxD	137.91	31	1.82	1
(2): A+B+C+D+BxC	85.37	31		
(2)+AxB	56.48	27	28.89	4
(2)+AxC	77.85	27	7.52	4
(2)+AxD	61.03	27	24.34	4
(2)+BxD	83.80	30	1.57	1
(2)+CxD	83.55	30	1.82	1
(3):A+B+C+D+AxB+BxC	56.48	27		
(3)+AxC	52.38	23	4.10	4
(3)+AxD	32.14	23	24.34	4
(3)+BxD	54.92	26	1.56	1
(3)+CxD	54.67	26	1.81	1
(4): A+B+C+D+AxB+AxD+BxC	32.14	23		

may be adequately described in a spatial context by means of that log-linear model. The parameter estimates of the model (7.2) are presented in Table 7.10, to be able to characterize the five areas from the survey in terms of the mentioned variables.

Table 7.10. Estimates of main-effects and first-order interaction effect
parameters from Table 7.7 (with 'centred-effect' constraints).

Parameter	Magnitude of the effect	Standardised value	Parameter	Magnitude of the effect	Standardised value
λ^A (1)	.510	3.041	λ^{AB} (1,1)	.596	3.556
λ^A (2)	-.906	-3.161	λ^{AB} (2,1)	-.599	-2.091
λ^A (3)	1.064	7.630	λ^{AB} (3,1)	.039	.280
λ^A (4)	-.209	-.848	λ^{AB} (4,1)	.217	.880
λ^A (5)	-.458	-2.105	λ^{AB} (5,1)	-.253	-1.161
λ^B (1)	.206	1.893	λ^{AD} (1,1)	-.151	-.901
λ^C (1)	-.444	-4.071	λ^{AD} (2,1)	-.001	-.004
λ^D (1)	.550	5.046	λ^{AD} (3,1)	-.262	-1.880
λ^{BC} (1,1)	-.486	-4.453	λ^{AD} (4,1)	.671	2.721
			λ^{AD} (5,1)	-.256	-1.178

The main effect parameters and the first-order interaction effect parameters
from Table 7.10 are defined as deviations of the overall mean-effect λ, which
is equal to 1.467. When analysing the results in the table with the fitting
procedure of the data in Table 7.7, the two variables 'type of boat' and
'distance to the home address' both differ in space. Finally, the type of
boat is also mutually related to the place to spend the night. The other
first-order interaction effects (viz. AxC, BxD and CxD), do not give a statistical significant decrease in deviance value.

An interpretation of the parameter values from Table 7.10 will be presented
in Table 7.11.

The interaction effect between the variables 'type of boat' and 'place to
spend the night' (i.e. the first-order interaction effect between the variables B and C) shows that the number of recreationers with a motorized boat
who spend the night at home (i.e. B=1, and C=1) is moderately below the average pattern in the survey.

The five areas from the survey in the Biesbosch area have been characterized
now in terms of the age and the educational level of the recreationers, as
well as in terms of the type of boat they use and the distance to their home
address. An explanatory analysis will be presented below concerning the place
where the daily purchases will be made by the recreationers in the Biesbosch
area. The response variable to be explained is the area where those purchases
will be made; it is classified into two distinct classes, viz. either or not
buying the daily purchases in the Biesbosch area. The explanatory variables
are 'area of visit', 'distance to the home address', and 'duration of staying
in the Biesbosch area'. The variables are cross-classified in a four-dimensional contingency table with respect to the response variable. This table
with size 2x5x2x3 is given in Table 7.12, and the explanatory variables are
denoted by A,B, and C successively.

Table 7.11. Spatial characterization of the recreationers of the survey,
according to the variables 'type of boat', 'place to spend the
night', and 'the distance between the home-address and the
Biesbosch area'.

Area	Characterization of the interaction-effect AxB
Rietplaat	The number of recreationers with a motorized boat is moderately above the average with respect to the pattern in the survey.
Honderd-dertig	The number of recreationers of the survey in this area with an unmotorized boat is moderately above the average pattern. One should take into account that in this area the absolute number of unmotorized boats is even larger than the number of motorized boats, which does not occur in the other areas of the survey.
Keesjes Killeke	The number of motorized boats in this area shows a pattern which is analogous to the distribution of the boats in the whole survey.
Merwelanden	The number of motorized boats in this area is slightly above the average pattern of the survey.
Zuid-hollandse Biesbosch	The number of unmotorized boats in this area is slightly above the average pattern of the survey.
Area	Characterization of the interaction-effect AxD
Rietplaat	The number of recreationers visiting this area, who live at a distance of less than 30 kilometers from the home address is moderately below the average pattern in the survey.
Honderdder-tig	The distance between the home address and the Biesbosch area for the respondents out of this area is just around the average pattern of the survey. The number of respondents who live close to the Biesbosch (i.e. less than 30 kilometers) is about 2.5 times more than the number of respondents who live further away.
Keesjes Killeke	Same conclusion as with respect to the area Rietplaat.
Merwelanden	Figure 7.1(d) already mentioned that this area has a much stronger local function than the other areas, because at least 50 percent of the recreationers live closer than 10 kilometers from the Biesbosch area. This conclusion is also supported by the strong interaction-effect between the variables A and D for this area.
Zuid-hollandse Biesbosch	Same conclusion as with respect to the area Rietplaat.

Table 7.12. Four dimensional contingency table on the daily purchases, according to the variables 'area of visit', 'distance to the home address', and 'number of days of staying'.

A	C = 1 B = 1					C = 1 B = 2					C = 2 B = 1					C = 2 B = 2					C = 3 B = 1					C = 3 B = 2				
	1	2	3	4	5	1	2	3	4	5	1	2	3	4	5	1	2	3	4	5	1	2	3	4	5	1	2	3	4	5
X=1	6	1	9	13	1	1	0	0	0	0	39	7	34	34	7	4	1	7	2	3	20	1	12	6	5	12	1	11	2	2
X=2	1	1	12	0	0	7	1	4	1	4	6	0	10	5	2	10	2	21	2	1	3	0	8	0	0	2	0	5	0	0
Total	7	2	21	13	1	8	1	4	1	4	45	7	44	39	9	14	3	28	4	4	23	1	20	6	5	14	1	16	2	2

X = 1: the daily purchases will be bought in the Biesbosch area;
X = 2: the daily purchases will be bought outside the Biesbosch area.

A = area: 1 = Rietplaat;
2 = Honderddertig;
3 = Keesjes Killeke;
4 = Merwelanden;
5 = Zuidhollandse Biesbosch

B = distance to home address:
1 = less than 30 kilometers;
2 = more than 30 kilometers.
C = number of days of staying in the Biesbosch.
1 = one day;
2 = a weekend, two or three days;
3 = at least four days.

A logistic regression model - in case of a two-category response variable - is presented in equation (5.26), and it has been rewritten as a linear logit model in equation (5.28). The information from Table 7.12 will be analysed by making use of a linear logit model, with variable X as the response variable to be related to the (interaction) effects between the variables A, B, and C. The linear logit model with all main effects included then becomes:

$$\log_e\left(\frac{p_{ijk}}{1 - p_{ijk}}\right) = \omega + \omega^A(i) + \omega^B(j) + \omega^C(k), \qquad \begin{array}{l} i = 1,..,5; \ j = 1,2; \\ k = 1,2,3. \end{array} \qquad (7.3)$$

in which p_{ijk} is the proportion of the recreationers in category i,j and k who buy the daily purchases in the Biesbosch area. The G^2-value of this model with all main effects is equal to 40.02 with degrees of freedom equal to 22. Model estimation with the GLIM program, and the model selection procedures discussed in subsection 5.3.4, may provide the relevant interaction effects to explain the daily purchases bought either in or outside the Biesbosch area.

Aitkin (1979; 1980) has introduced the simultaneous test procedure to be a model selection procedure which explicitly includes the possibility that one may be misled by assigning significance to parameters to what is merely random variation.

Consider the linear logit model in equation (7.3), with the null hypothesis that all three first-order interaction effects and the one second-order interaction effects are zero. When a type one error rate α of 0.10 is established, the overall type one error rate γ for that model with all main effects included is given by $\gamma = 1-(1-\alpha)^4 = 1-(0.90)^4 = .3439$. This means that the probability of at least one incorrect rejection of the null-hypothesis (i.e. all first- order and second-order interaction effects zero) is 34%. The overall type one error rate γ of .34 is within the range of (.25,.50) which was recommended by Aitkin.

With the level of significance α equal to 0.10 the critical values of the pooled effects are the following:

$\gamma_2 = 1-(1-\alpha)^3 = 1-.729 = .271$,

$\gamma_3 = 1-(1-\alpha) = .10$,

in which γ_3 is the level of significance to test the second-order interaction effect, and γ_2 is the level of significance to test the first-order interaction effects.

The deviance value of the second-order interaction effect AxBxC is 3.673 which is not significant because the critical chi-squared value with 8 degrees of freedom, and a 0.10 level of significance, is equal to 13.4.

However, the significance level of the pooled first-order and second-order interaction effects may be determined by:

$\gamma_{2,3} = 1-(1-\gamma_2)(1-\gamma_3) = 0.3439$.

The chi-squared critical value with 22 degrees of freedom and a .3439 level of significance equals to 24.35, and the pooled first-order and second-order interaction effects are significant because the G^2-value is equal to 40.02.

Table 7.13 shows the results of the stepwise selection when applied to the data of Table 7.12.

Table 7.13 shows the goodness-of-fit statistics of six sets of linear logit models, each of them starting from the bottom-up with the saturated model. The aim of the approach is to eliminate those terms of the saturated model until the critical chi-squared value (which is 24.35 in this case) is exceeded. The models selected from the six stages are successively A+B+C+AxB+AxC, A+B+C+AxB+BxC, A+B+C+AxB+AxC, A+B+C+AxC+BxC, A+B+C+BxC, and A+B+C+BxC.

The model selected by the simultaneous test procedure of Table 7.13 is A+B+C+ BxC, with a goodness-of-fit statistic equal to 20.02 and 20 degrees of freedom. The interaction effects included to the base model in equation (7.3) is the diversity with respect to the distance to the home address and the number of days of staying (the BxC term). The parameter values of the logit model selected, which have been obtained from the GLIM procedure, are presented in Table 7.14.

Table 7.13. Stepwise selection applied to the 2x5x2x3 table from Table 7.12.

Model	G^2	d.f	Parameter	Deviance	d.f.
(1)A+B+C	40.02	22			
A+B+C+AxB	37.93	18	AxB	2.09	4
A+B+C+AxB+AxC	24.30	10	AxC	13.63	8
A+B+C+AxB+AxC+BxC	3.673	8	BxC	20.627	2
A+B+C+AxB+AxC+BxC+AxBxC	0.0	0	AxBxC	3.673	8
(2)A+B+C	40.02	22			
A+B+C+AxB	37.93	18	AxB	2.09	4
A+B+C+AxB+BxC	18.76	16	BxC	19.17	2
A+B+C+AxB+BxC+AxC	3.673	8	AxC	15.087	8
A+B+C+AxB+BxC+AxC+AxBxC	0.0	0	AxBxC	3.673	8
(3)A+B+C	40.02	22			
A+B+C+AxC	29.12	14	AxC	10.90	8
A+B+C+AxC+AxB	24.30	10	AxB	4.82	4
A+B+C+AxC+AxB+BxC	3.673	8	BxC	20.627	8
A+B+C+AxC+AxB+BxC+AxBxC	0.0	0	AxBxC	3.673	8
(4)A+B+C	40.02	22			
A+B+C+AxC	29.12	14	AxC	10.90	8
A+B+C+AxC+BxC	7.176	12	BxC	21.944	2
A+B+C+AxC+BxC+AxB	3.673	8	AxB	3.503	4
A+B+C+AxC+BxC+AxB+AxBxC	0.0	0	AxBxC	3.673	8
(5)A+B+C	40.02	22			
A+B+C+BxC	20.20	20	BxC	19.82	2
A+B+C+BxC+AxB	18.76	16	AxB	1.44	4
A+B+C+BxC+AxB+AxC	3.673	8	AxC	15.087	8
A+B+C+BxC+AxB+AxC+AxBxC	0.0	0	AxBxC	3.673	8
(6)A+B+C	40.02	22			
A+B+C+BxC	20.20	20	BxC	19.82	2
A+B+C+BxC+AxC	7.176	12	AxC	13.024	8
A+B+C+BxC+AxC+AxB	3.673	8	AxB	3.503	4
A+B+C+BxB+AxC+AxB+AxBxC	0.0	0	AxBxC	3.673	8

Table 7.14. Parameter estimates of the logit model fitted to Table 7.12.

Parameter	Estimate	Standard error
grand mean	1.114	0.4332
ω^A (2)	.2134	.7531
ω^A (3)	-.9837	.3285
ω^A (4)	.6653	.4956
ω^A (5)	.5580	.6741
ω^B (2)	-4.024	1.117
ω^C (2)	.7679	.4165
ω^C (3)	.6138	.4938
ω^{BC} (2,2)	1.741	1.174
ω^{BC} (2,3)	4.158	1.252

The main effect parameters of variable A in Table 7.14 show the spatial di-
diversity concerning the place where the daily purchases will be bought. The
recreationers who visit the area Merwelanden especially buy their daily purchases in the Biesbosch area, and the recreationers who visit the areas Honderddertig and Zuidhollandse Biesbosch show a weak tendency to buy the daily
purchases in the Biesbosch area.

The main effect of the variable B shows the diversity of the distance categories with respect to variable X. The negative value of this parameter in
Table 7.14 (which is related to the first category of that variable) strongly
indicates that the recreationers who live more than 30 kilometers from the
Biesbosch area buy their daily purchases outside the Biesbosch.

The first-order interaction effect BxC has - in case of the second category
of variable B - positive estimate values for the categories 2 and 3 from
variable C. This shows that the recreationers who live more than 30 kilometers from the home address buy their daily purchases in the Biesbosch area
when they stay there for a longer period.

7.3. THE MULTIVARIATE NATURE OF OUTDOOR RECREATION

7.3.1. Introduction

Various phenomena concerning outdoor recreation in the Biesbosch area have
been analysed in the previous section. Some examples are inter alia a spatial
characterization of recreationers in terms of the age and the level of education, a spatial characterization with respect to the type of boat and the
distance to the home address, as well as the percentage of recreationers who
buy the daily purchases in the Biesbosch area in terms of the duration of
their stay and the distance to the home address.

Two multivariate approaches have been discussed in subsections 5.4.2 and
5.4.3, i.e. multidimensional scaling for ordinal information, and the HOMALS
procedure for nominal information. Both procedures aim to derive a spatial
configuration of a system which is multivariate in nature and denoted at a
non-metric scale.

The first part of this section shows a multivariate analysis of five variables by making use of the HOMALS program, with information obtained from the
survey (see also subsection 7.3.2). This analysis shows the spatial diversity
with respect to potentially relevant phenomena to come to the Biesbosch
area.

The second part of this section focuses upon a multivariate analysis with
respect to type of boat, type of bank and type of recreational activities.

This analysis makes use of a multidimensional scaling procedure (see also subsection 7.3.3).

7.3.2. Homogeneous scaling of outdoor recreation

The relevance of either environmental phenomena (e.g., character of land-scape) or of the availability of recreational facilities, to select an area for recreational activities, will be analysed simultaneously in this subsection with the HOMALS program. The five variables, with three categories each, all reflect the relevance to come to the Biesbosch area, and will be descri-bed below.

Variable 1 reflects the distance to the residence as being relevant to visit the Biesbosch area. The second variable reflects the relevance of being in a position to sail, fish, or surf in an area such as the Biesbosch. Variable 3 are the recreational facilities such as, for example, beaches and landing-stages, as being an important reason to visit the Biesbosch area. Variables 2 and 3 especially refer to the relevance of recreational facilities to spend leisure time in the study area, and the variables 4 and 5 are the environmen-tal variables, which may also be a relevant motivation for recreationers in their selection of the Biesbosch area. Variable 4 refers to the character of landscape, and variable 5 refers to the ecological characteristics of the area, such as the presence of plants and animals.

The three categories of these five variables are classified in terms of an unimportant (category 1), a slightly important (category 2), and an important (category 3) reason to spend the leisure time in the Biesbosch area.

The HOMALS program is used for each of the five subregions in the survey. Table 7.15 shows the discrimination measures of the five variables described above.

The eigenvalues which are shown in Table 7.15 for each dimension represent the homogeneity of the data with respect to that dimension, in such a way that the higher the eigenvalue, the more homogeneous are the data. The eigen-value (or homogeneity) is equal to the average value of the discrimination measures for that dimension. The higher a discrimination measure of a varia-ble, the better a variable discriminates on that dimension, and the more important that variable is for the solution. A large discrimination measure also represents a large dispersion of the categories of that variable. Two general characteristics of the discrimination measures are:

- when two variables have the same distribution with respect to the category scores, the discrimination measures of these variables then are close to each other;

- when the distribution of the category scores of some variable fits well
with respect to the marginal distribution of all variables, the discrimination measure of that variable is close to the origin.

Table 7.15. Discrimination measures of the HOMALS analysis concerning the
relevance of the five phenomena.

Table 7.15a. Discrimination measures
of the HOMALS analysis
for the area Rietplaat.

Variables	Dimension 1	Dimension 2
1	. 252	. 043
2	. 335	. 605
3	. 124	. 549
4	. 506	. 003
5	. 452	. 069
Eigenvalue	.3337	.2540

Table 7.15b. Discrimination measures
of the HOMALS analysis
for the area Honderd-
dertig.

Variables	Dimension 1	Dimension 2
1	. 713	. 137
2	. 797	. 066
3	. 467	. 866
4	. 000	. 004
5	. 009	. 471
Eigenvalue	.3973	.3088

Table 7.15c. Discrimination measures
of the HOMALS analysis
for the area Keesjes
Killeke.

Variables	Dimension 1	Dimension 2
1	. 063	. 543
2	. 273	. 193
3	. 229	. 362
4	. 611	. 004
5	. 442	. 276
Eigenvalue	.3234	.2757

Table 7.15d. Discrimination measures
of the HOMALS analysis
for the area Merwelanden.

Variables	Dimension 1	Dimension 2
1	. 013	. 149
2	. 215	. 308
3	. 334	. 544
4	. 515	. 469
5	. 662	. 011
Eigenvalue	.3476	.2964

Table 7.15e. Discrimination measures
of the HOMALS analysis
for the area Zuidhollandse
Biesbosch.

Variables	Dimension 1	Dimension 2
1	. 544	. 178
2	. 361	. 119
3	. 541	. 385
4	. 493	. 616
5	. 610	. 591
Eigenvalue	.5100	.3777

Table 7.15f. Discrimination measures
of the HOMALS analysis
for the Biesbosch area.

Variables	Dimension 1	Dimension 2
1	. 071	. 265
2	. 296	. 182
3	. 091	. 389
4	. 513	. 231
5	. 591	. 218
Eigenvalue	.3122	.2569

The discrimination measures for the area Rietplaat are presented in Table 7.15(a). The variables 4 and 5 have high discrimination measures on the first dimension and the variables 2 and 3 have high discrimination measures on the second dimension. The two variables concerning the relevance of recreational facilities in this area (viz. the relevance for recreationers to be in a position to sail or to fish, and the relevance of the presence of recreational facilities) show a similar pattern. The major part of the respondents replied that these two variables are important reasons to visit the area (78% and 68% successively). This conclusion with respect to the relevance of recreational facilities to come to the area Rietplaat is in agreement with Section 7.2 where this area was characterized as one with a lot of recreational facilities. The two variables concerning the relevance of the environmental phenomena (viz. the character of landscape and the presence of plants and animals) also show similar patterns with respect to the discrimination measures.

Variable 1 (the relevance of staying close to the home address) is an outlier in this area with respect to the five variables, with about 27% of the respondents replying that this variable is only slightly important.

The discrimination measures for the area Honderddertig are presented in Table 7.15(b). The variables 1 and 2 have high discrimination measures on the first dimension, and the variable 3 has a high discrimination measure on the second dimension. The HOMALS-procedure discriminates the relevance of environmental phenomena with that of the recreational facilities. The response to the variables 1 and 2 show similar patterns in this area, with about 20% response to category 1 (unimportant as a reason) for both variables. The variables 4 and 5 show a distinct pattern from the other variables, because no respondent from the survey replied that any of these environmental phenomena is unimportant to come to the Biesbosch.

All respondents indicated that the nature of landscape is a relevant reason to come to this area. A distinct pattern is shown here for variable 3, with a 56% response level indicating that it is an unimportant reason to come to the area because of recreational facilities. This conclusion concerning the relevance of environmental phenomena is in agreement with the conclusion already mentioned in the previous section, that this area has only a minor level of recreational facilities.

Table 7.15(c) shows the discrimination measures for the area Keesjes Killeke. The variables 4 and 5 have high discrimination measures on the first dimension, and the variables 1 and 3 have high discrimination measures with respect to the second dimension. The discrimination measures show that the variables 1 and 3 have similar patterns. At least 25% of the response on these variables mentioned them as unimportant reasons to come to this area. Variables 4 and 5 are also clustered, and less than 10% of the respondents replied that any of these two environmental criteria is an unimportant reason to go to the area Keesjes Killeke.

Table 7.15(d) shows the discrimination measures for the area <u>Merwelanden</u>. The discrimination measure for the first dimension is high on the variables 4 and 5, and for the second dimension it is high on the variables 3 and 4. Variable 1 has small discrimination measures on both dimensions and this variable is therefore only slightly important for the solution. The main difference between variables 4 and 5 is due to the response to category 2, because only 8% of the response replied that the character of landscape is slightly important, while 32% of the response replied that the presence of plants and animals is a slightly important reason to come to the area Merwelanden.

Table 7.15(e) shows the discrimination measures for the area <u>Zuidhollandse Biesbosch</u>. The results in this table show that all variables fit well in the first dimension, and that the variables 4 and 5 also have a reasonable fit in the second dimension. 'Distance to the home address' and 'the availability of recreational facilities' are less important items than in the other areas, except for the area Honderddertig with a 56% reply that the availability of recreational facilities is an important reason. However, in the area Zuidhollandse Biesbosch some 35% of the response mentioned that variable 1 or variable 3 is an unimportant item to come to that area.

The discrimination measures for the <u>Biesbosch area</u> is shown in Table 7.15(f), with a total sample size of 378. The discrimination measures for the first dimension is high on the variables 4 and 5, and for the second dimension it is high on the variables 1 and 3. The category quantifications of the variables 4 and 5 show similar patterns, as well as with respect to the variables 1 and 3. Some 25% of the response replied that variable 1 is an unimportant reason and 20% replied that variable 3 is an unimportant reason to come to the Biesbosch area. About 50% mentioned these variables to be important. However, some 85% and 65% of the respondents mentioned that variables 4 and 5 successively are an important item to spend the leisure time in the Biesbosch.

Environmental phenomena such as the nature of landscape and the existence of plants and animals are important in all subareas of the Biesbosch. A spatial difference in the Biesbosch area exists especially with respect to the relevance of the recreational facilities for recreationers to come to the Biesbosch.

7.3.3. Multidimensional scaling and outdoor recreation

The relevance of environmental phenomena and the availability of recreational facilities for recreationers is described in the previous subsection by making use of the HOMALS program. The results showed that especially the relevance of recreational facilities differs with respect to the five Biesbosch areas.

The recreational patterns in the Biesbosch area will therefore be further discussed in this subsection in terms of a spatially disaggregated analysis of recreational activities, type of bank, and type of boat. The recreational patterns are subdivided into three main groups with 10 variables in all, viz. type of boat (with 2 variables), type of bank (with 4 variables), and type of activities (with 4 variables). The Biesbosch area has been subdivided into 11 subregions to be able to present a spatially disaggregated analysis of recreational activities. The variables and the areas are given below.

Variables

Type of boat   : variables X_1 and X_2;
Type of bank   : variables X_3, X_4, X_5 and X_6;
Type of activities: variables X_7, X_8, X_9 and X_{10}.
X_1 = motorized boats (motor boats, open boats with a motor);
X_2 = unmotorized boats (sailing boats without a motor, canoes and surfing boards);
X_3 = willow-trees and holm;
X_4 = reed vegetation and saltings;
X_5 = landing-stages;
X_6 = beaches;
X_7 = activities from the boat such as fishing;
X_8 = activities in water such as swimming;
X_9 = 'quiet' activities on the beach (sun-bathe, walking, e.g.);
X_{10} = 'unquiet' activities on the beach (camp out, playing football, e.g.).

Areas

 1 = Brabant Biesbosch, western part;
 2 = Brabant Biesbosch, south-western part;
 3 = Brabant Biesbosch, south-eastern part;
 4 = Brabant Biesbosch, eastern part;
 5 = Brabant Biesbosch, north-eastern part;
 6 = Brabant Biesbosch, northern part;
 7 = Brabant Biesbosch, rush-saltings area;
 8 = Brabant Biesbosch, Heenplaat;
 9 = Dordrecht Biesbosch area;
10 = Sliedrecht Biesbosch area (western part);
11 = Sliedrecht Biesbosch area (eastern part).

Table 7.16. Two dimensional metric representation of the variables by means of MDS.

Area = 1

Variables	Dimension 1	Dimension 2
1	-.2604	-.1721
2	-.0284	.0953
3	-.6472	-1.000
4	-.4789	-.5408
5	.4613	-.4350
6	-.0140	.4028
7	-.1501	.0488
8	-.0570	.2129
9	-.7183	-.3023
10	.6315	.1450

Area = 2

Variables	Dimension 1	Dimension 2
1	-.0438	.0171
2	.2280	-.1393
3	-.2206	.1199
4	-.1697	.0917
5	.2240	-.1557
6	.2818	.0118
7	-.1889	.1014
8	-.1307	.0682
9	.2343	-.1461
10	.2282	-.1605

Area = 3

Variables	Dimension 1	Dimension 2
1	.2307	.1618
2	.2308	.1621
3	.2313	.1633
4	.2520	-.1752
5	.2312	.1632
6	-.0661	.0785
7	.2313	.1607
8	.2310	.1618
9	-.0700	.0862
10	-.1807	.3573

Area = 4

Variables	Dimension 1	Dimension 2
1	-.2247	-.0885
2	-.1847	.1299
3	-.6609	.0429
4	-.2717	-.0634
5	-.0649	-.5756
6	.2998	.0429
7	-.3577	.1287
8	-.1040	.1195
9	.1368	-.1401
10	.1308	.2849

Area = 5

Variables	Dimension 1	Dimension 2
1	-.1492	.2062
2	-.3556	.3064
3	.4124	-.2636
4	-.3610	.2332
5	-.2776	.4692
6	-.3685	.3410
7	-.1970	.0500
8	.3271	-.2366
9	-.1051	.5732
10	.0420	.6342

Area = 6

Variables	Dimension 1	Dimension 2
1	.1288	-.1231
2	.1202	-.4387
3	.1720	.0387
4	-.0036	-.6488
5	-.1508	-.0402
6	.2640	.1101
7	-.1260	.3043
8	.1450	-.4621
9	.1824	.1771
10	.2323	.1402

Table 7.16 (continued).

Area = 7

Variables	Dimension 1	Dimension 2
1	-.1130	-.0160
2	.1182	-.1058
3	-.1060	.0305
4	.1157	-.0677
5	.2599	.1129
6	.4471	-.0730
7	-.4136	-.0527
8	-.1343	-.0736
9	-.0632	.0283
10	.2265	.2953

Area = 8

Variables	Dimension 1	Dimension 2
1	.1118	-.2615
2	-.3321	-.0982
3	.1852	.3487
4	.1040	.6511
5	.4914	-.0442
6	-.3511	-.0866
7	.1495	.3364
8	.1025	-.3709
9	.1858	-.2762
10	-.5199	-.0136

Area = 9

Variables	Dimension 1	Dimension 2
1	.4308	.0216
2	-.7635	-.0157
3	.4317	.0252
4	-.1287	-.0078
5	-.2697	-1.000
6	-.2348	-.7603
7	.6142	-.0111
8	-.5129	-.0110
9	-.3831	-.4092
10	.0012	-.1915

Area = 10

Variables	Dimension 1	Dimension 2
1	-.1805	.2737
2	-.1284	-.1300
3	-.4027	-.0826
4	-.0615	.4012
5	-.3127	-.1768
6	-.5232	-.0917
7	-.1205	-.0670
8	-.0823	.4654
9	-.0740	.4158
10	.3212	-.1737

Area = 11

Variables	Dimension 1	Dimension 2
1	-.1284	-.1310
2	-.2590	-.1816
3	-.1743	-.1523
4	-.6711	-.3099
5	.0885	-.3127
6	-.0535	-.1059
7	-.1381	-.1322
8	-.1582	.5934
9	.5477	.0469
10	.5282	.0744

The aim of the analysis is to investigate the spatial patterns of the variables, and the originally non-metric information is therefore transformed in a
two-dimensional Euclidean space. A metric representation of the recreational
profiles (type of boat, use of natural resources such as willow-trees and
reed-vegetation, and recreational activities) is achieved by means of a
multidimensional scaling procedure.

The results of the MDS-procedure for each of the 11 regions are presented in
Table 7.16. This table shows that the major part of the variables appear in
two or three quadrants, which indicates that similarity in pattern concerning
the variables can be obtained. Table 7.17 shows the results of the MDS-procedure with the variables depicted in terms of the quadrants in a two-dimensional Euclidean space.

Table 7.17. Two-dimensional representation of the recreational profiles.

Area	1st quadrant	2nd quadrant	3rd quadrant	4th quadrant
1	10	2,6,7,8	1,3,4,9	5
2	6	1,3,4,7,8	---	2,5,9,10
3	1,2,3,5,7,8	6,9,10	---	4
4	6,10	2,3,7,8	1,4,5	9
5	10	1,2,4,5,6,7,9	---	3,8
6	3,6,9,10	7	4,5	1,2,8
7	5,10	3,9	1,7,8	2,4,6
8	3,4,7	---	2,6,10	1,5,8,9
9	1,3	---	2,4,5,6,8,9	7,10
10	---	1,4,8,9	2,3,5,6,7	10
11	9,10	8	1,2,3,4,6,7	5

The variables depicted in some quadrant show similar patterns with respect to
the grids in the area.

The results from the MDS-procedure show that the variables X_1 and X_2 (motorized and unmotorized boats respectively) have dissimilar patterns with respect to the other variables in 7 out of the 11 regions (viz. areas 1,2,4,7,
8,9 and 10).

The results of the MDS-procedure will be discussed briefly for each of the 11
regions.

The western part of the Biesbosch (area 1) shows a distinction concerning
the type of boat with respect to the type of bank used by the boats, because
the motorized boats have a similar pattern with the banks willow-trees and
reed-vegetation, while the unmotorized boats have a similar pattern with the
use of beaches. A distinct pattern for the variables X_5 (landing-stages) and
$X10$ ('unquiet' activities) may be concluded because they are represented in
different quadrants.

The south-western part of the Biesbosch (area 2) also shows a distinction in pattern with respect to the motorized and unmotorized boats: the motorized boats make use of willow-trees and reed-vegetation, and the unmotorized boats make use of landing-stages. With respect to the type of activities in this area, the motorized boats show similar patterns with activities such as fishing and swimming, and the unmotorized boats show similar patterns with activities on the beach (the variables X_9 and X_{10}).

The main characteristic of the south-eastern part of the Biesbosch, in comparison with the other areas, is the similarity in pattern with respect to the two types of boat, the willow-trees and landing-stages as the type of bank, and the recreational activities such as fishing and swimming.

The results of MDS for the eastern part of the Biesbosch show the similarity between motorized boats and two types of bank, viz. reed-vegetation and landing-stages, while the unmotorized boats show a similarity with respect to the willow-trees and activities such as fishing and swimming.

The north-eastern part of the Biesbosch (area 5) is mainly characterized by its similarity in pattern concerning the two types of boat, the three types of bank (willow-trees excluded), and recreational activities such as fishing and sun-bathe. Activities in water such as swimming will mainly be done with one type of bank, viz. willow-trees and holm.

The main characteristic of the northern part of the Biesbosch (area 6), in comparison with the other areas, is the similarity in pattern of the two types of boat and the recreational activities in water such as swimming.

The rush-saltings area of the Biesbosch (area 7) shows a similarity between the use of motorized boats and the activities from the boats such as fishing, and activities in water such as swimming. The unmotorized boats in this area especially make use of two types of bank, viz. reed vegetation and beaches.

The area Heenplaat (area 8) has a distinctive pattern with motorized and unmotorized boats. The motorized boats on the one hand especially make use of landing-stages, and the recreationers make use of recreational activities in water such as swimming, and 'quiet' activities on the beach. The unmotorized boats in this area make use of beaches as the type of bank and the recreationers participate in 'unquiet' recreational activities.

Area 9 is the Dordrecht Biesbosch area , with motorized boats making use of willow-trees and holm as the type of bank. A very similar pattern in this area can be seen with the unmotorized boats and the recreational activities in water such as swimming.

The Sliedrecht Biesbosch area has been subdivided into the western and the eastern part. The recreationers in the western part of the Sliedrecht Biesbosch area show a distinctive pattern with the use of motorized and unmotorized boats. The motorized boats especially make use of reed vegetation as the type of bank, with activities in water such as swimming, as well as 'quiet' activities on the beach. The unmotorized boats on the other hand, make use of reed vegetation and landing-stages, with activities from the boat such as fishing. No distinctive pattern can be traced out in the <u>eastern part of the Sliedrecht Biesbosch area</u> with respect to the motorized and unmotorized boats.

7.4. PATH ANALYSIS AND RECREATIONAL ACTIVITIES

The recreational activity patterns in the Biesbosch area have been discussed in the previous subsection in terms of the type of boat used by recreation-ers, and the type of bank where the boats will be laid. The analysis made use of a multidimensional scaling procedure to represent 10 variables in a two-dimensional space and shows the similarity or dissimilarity patterns of the variables with respect to recreational activities. The Biesbosch region was disaggregated into 11 regions to show the spatial diversity of the variables. The cause-and-effect relationships between type of boat, type of bank, and type of recreational activities will be analysed in this section for the 10 variables mentioned in the previous subsection. Such relationships may be analysed by means of path analysis (see also subsection 5.2.4 where systems were described in terms of path analysis).

Two types of non-recursive path models will be presented in this section, viz.:

(a) the relationship between the type of boat, and the type of bank used by the boats;

(b) the relationship with respect to recreational activities in terms of the type of boat, and the type of bank used.

The first model describes the difference of cause-and-effect relationships concerning the use of motorized and unmotorized boats for each of the four types of bank, and these relationships are visualized in Figure 7.2.

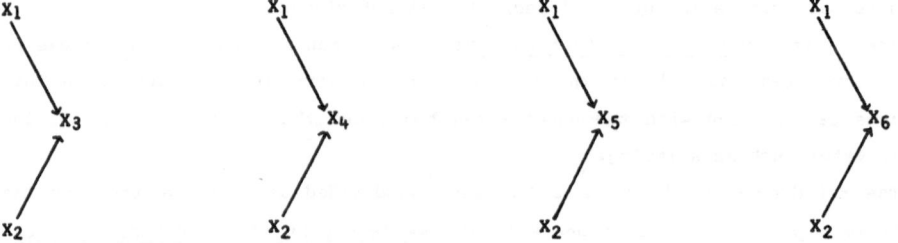

Figure 7.2. Cause-and-effect relationships of type of boat and type of bank.

The second model describes the use of motorized and unmotorized boa   r
each of the four types of bank with respect to four types of recrea   l
activities. Figure 7.3 shows the four relationships of each of the r   au-
tional activities with respect to the use of willow-trees and holm (va   ble
X_3) and motorized boats (variable X_1) or unmotorized boats (variable X_2

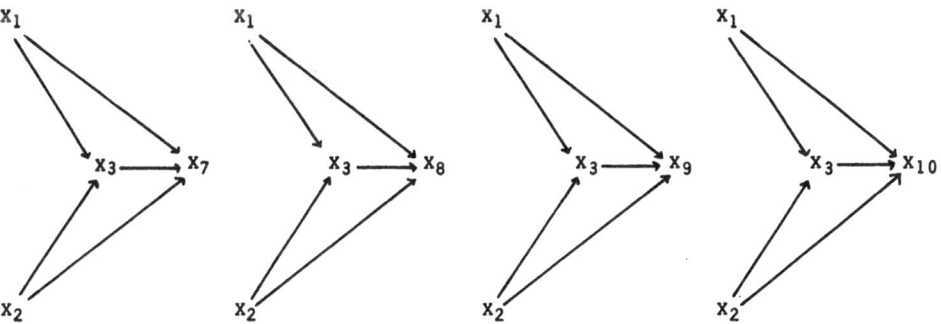

Figure 7.3. Cause-and-effect relationships of type of boat, use of willow-
trees and four types of recreational activities.

The path models for the variables X_4, X_5, and X_6 are specified analogously to
the one presented in Figure 7.3 for the variable X_3.

The path coefficients of the models from Table 7.2 and 7.3 are presented in
Table 7.18. This table is subdivided into four tables according to the four
types of bank. The path coefficients of the motorized and unmotorized boats
with respect to the use of willow-trees are presented in the first part of
Table 7.18(a), and the path coefficients of the other types of bank are presented in Table 7.18 (b)-(d) successively. The path coefficients p_{ij} are
used to identify the path to variable X_i from variable X_j. The coefficients p_i are used to identify the path from the error term, and is defined as the square root of the proportion of variance explained by that
error term.

The path coefficients of the relationships in Figure 7.2 are discussed in
Table 7.19.

Table 7.18. Path analysis according to types of activities, banks and boats.

Table 7.18a. Path analysis of the variable X_3 according to four types of activities.

Area	P_{31}	P_{32}	P_3	P_{71}	P_{72}	P_{73}	P_7	P_{81}	P_{82}	P_{83}	P_8	P_{91}	P_{92}	P_{93}	P_9	P_{101}	P_{102}	P_{103}	P_{10}
1	.76	-.09	.68	.46	.06	.32	.67	.27	.03	.32	.83	.49	.17	.04	.80	.38	-.02	-.17	.96
2	1.12	-.46	.61	-.64	.63	.70	.79	.54	.54	-.18	.37	.94	.12	-.27	.50	.48	.58	-.23	.47
3	-.14	.58	.84	.22	.26	.45	.64	.87	.02	-.04	.48	.82	.17	-.21	.45	.56	.20	-.32	.74
4	.06	.72	.65	-.29	.68	.17	.71	.58	.26	-.22	.76	.40	.35	-.34	-.86	-.00	.57	-.43	.92
5	.79	-.35	.74	.11	.12	.25	.92	.19	.06	.27	.90	.64	-.06	-.24	.86	-.28	.66	.13	.83
6	.92	-.03	.39	-.05	.04	.44	.91	1.84	-.12	-1.33	.53	1.43	-.03	-.63	.47	.99	-.14	-.18	.57
7	.90	-.13	.56	.91	.03	-.37	.75	.92	.00	-.31	.72	.10	.40	.54	.48	-.24	-.03	.48	.73
8	.43	-.17	.92	-.29	.14	.50	.88	.62	.24	-.14	.58	.75	.15	.13	.47	.07	.93	-.14	.26
9	.87	.06	.44	.92	.55	-.66	.64	1.34	.22	-.87	.64	-.21	-.04	.35	.97	-.24	.01	1.18	.18
10	.80	.16	.51	.42	.50	-.00	.66	.57	.54	-.60	.82	.35	.13	.41	.61	.16	-.09	-.12	.98
11	.36	.34	.77	-.68	-.16	.40	.49	.61	-.62	.15	.82	.63	.51	-.53	.59	.38	.50	-.54	.78
All	.70	.07	.73	.18	.20	.18	.90	.82	-.02	-.32	.76	.87	-.03	-.36	.74	.27	.31	-.02	.88

Table 7.18. (continued).

Table 7.18b. Path analysis of the variable X₄ according to four types of activities.

Area	P_{41}	P_{42}	P_4	P_{71}	P_{72}	P_{74}	P_7	P_{81}	P_{82}	P_{84}	P_8	P_{91}	P_{92}	P_{94}	P_9	P_{101}	P_{102}	P_{104}	P_{10}
1	.84	-.10	.58	.64	.04	.07	.70	.44	.01	.07	.86	.35	.19	.21	.80	.49	-.04	-.30	.95
2	.74	-.07	-.72	-.37	.35	.69	.75	.35	.62	-.02	.39	.94	-.23	-.42	.44	.45	.66	-.31	.43
3	-.01	.35	.93	.16	.40	.32	.73	-.88	-.02	.03	.48	.85	.07	-.07	.48	.60	.06	-.14	.78
4	.60	-.14	-.72	-.60	.73	.54	.61	.72	.13	-.25	.75	.09	.04	.47	.82	.18	.31	-.35	.93
5	.22	.63	.62	-.37	.73	.64	.85	.28	-.36	.51	.86	.37	-.21	.37	.85	.87	-.10	-.40	.80
6	.24	.25	.92	.16	.06	-.15	.92	.55	.09	.25	.71	.88	.03	-.17	.51	-.26	.36	.80	.55
7	.58	.41	.45	.14	-.23	.74	.70	-.19	-.54	1.43	.36	.29	.13	.49	.52	1.11	.22	-.16	.55
8	.29	-.34	.94	-.23	.23	.53	.86	.68	.22	-.00	.60	.80	.13	.02	.49	.03	.93	-.07	.28
9	.70	-.05	.73	.10	.52	.34	.66	.17	.20	.60	.61	.11	-.02	-.03	.99	1.24	.06	-.65	.29
10	.58	.44	.56	.25	.38	.29	.64	-.27	.16	.64	.79	.75	.25	-.13	.64	-.14	-.27	.36	.97
11	1.08	-.48	.47	.01	.34	.75	.46	1.16	-.79	-.46	.80	1.34	-.07	-.83	.60	1.41	-.24	-1.14	.71
All	.36	.32	.83	.16	.06	.40	.85	.63	.04	-.09	.80	.67	.05	-.15	.78	.30	.36	-.13	.87

Table 7.18. (continued).

Table 7.18c. Path analysis of the variable x_5 according to four types of activities.

Area	P51	P52	P5	P71	P72	P75	P7	P81	P82	P85	P8	P91	P92	P95	P9	P101	P102	P105	P10
1	.32	-.07	.96	.74	.02	-.13	.69	.58	-.01	-.25	.82	.55	.16	-.09	.80	.06	.03	.59	.79
2	-.31	1.01	.62	.34	-.35	.65	.81	.29	.78	-.15	.38	.58	.41	-.16	.53	.22	.68	.01	.49
3	.65	-.15	.80	.15	.52	.01	.79	1.05	-.05	-.27	.43	.72	.08	-.04	.46	.07	.13	.82	.44
4	.89	-.06	.51	-.09	.79	-.21	.72	.98	.07	-.47	.73	1.21	.04	-.93	.75	.14	.25	-.19	.96
5	.39	.64	.40	.80	.82	-1.24	.79	.46	.07	-.16	.92	.38	-.08	.17	.88	.32	1.41	-1.25	.66
6	-.01	.16	.98	.35	-.00	.17	.91	.61	.16	-.05	.74	.85	-.04	.14	.52	.84	-.14	.03	.57
7	.69	.11	-.64	1.10	.16	-.76	.60	.87	.08	-.32	.71	.73	.36	-.22	.55	.20	-.16	.68	.57
8	.51	-.15	-.88	-.02	.03	-.10	.99	.80	.18	-.24	.56	.98	.08	-.34	.39	.14	.91	-.26	.17
9	-.05	.99	.18	.36	.20	.31	.70	.51	1.56	-1.40	.70	-.04	2.59	-2.62	.86	.79	.20	-.11	.55
10	-.01	.84	.55	.42	.72	-.26	.65	.10	.45	-.01	.87	.68	.23	-.04	.64	.07	-.18	.09	.99
11	.19	.57	.71	.92	.28	-.53	.44	.69	-.48	-.15	.82	.27	-.19	.91	.31	-.04	-.38	1.21	.24
All	.30	.58	.67	.40	.38	-.32	.88	.65	.13	-.21	.79	.63	.03	-.06	.78	.35	.50	-.32	.85

Table 7.18. (continued).

Table 7.18d. Path analysis of the variable X_6 according to four types of activities.

Area	P61	P62	P6	P71	P72	P76	P7	P81	P82	P86	P8	P91	P92	P96	P9	P101	P102	P106	P10
1	-.15	.76	.68	.64	.37	-.44	.63	.47	-.16	-.21	.85	.47	.43	-.35	.76	.25	-.02	.01	.97
2	.27	.65	.48	.48	1.12	-1.27	.66	.25	.41	.34	.36	-.46	-.18	.66	.43	.08	.33	.54	.41
3	.87	-.07	.54	.66	.48	-.58	.73	.49	.02	.45	.41	.93	.04	-.09	-.48	1.28	-.04	-.78	.66
4	.50	-.06	-.88	-.20	.80	-.16	.71	.38	.12	.38	.70	.15	.13	.46	-.80	-.20	.28	.35	.91
5	-.06	.85	.57	.33	.43	-.43	.91	.42	-.30	.31	.90	.49	-.59	.72	.78	-.13	.72	.72	.72
6	.57	-.07	.82	.54	.00	-.32	.89	.23	.20	.67	.50	-.73	-.00	.20	.51	.59	-.10	.43	.45
7	.46	.12	.83	.76	.13	-.39	.70	.67	.05	-.06	.73	.61	.34	-.09	.56	.56	.01	-.10	.78
8	.09	.92	.27	.03	1.18	-1.23	.94	.68	.29	-.08	.60	.77	-.24	-.40	-.48	-.08	.04	.99	.11
9	.11	.06	.98	.33	.50	.10	.70	.54	.14	.36	.66	.03	-.05	.50	.86	.80	.10	-.09	.55
10	.80	.19	.48	.69	.57	-.35	.64	.81	.61	-.89	.76	.71	.20	-.04	.64	.76	.06	-.86	.90
11	.71	.26	.45	.81	-.03	.02	.58	.28	-.71	.54	.79	.81	.47	-.52	.68	.79	.53	-.85	.80
All	.52	.11	.82	.41	.21	-.20	.89	.27	-.06	.62	.62	.39	-.05	.42	.71	.12	.29	.26	.85

Table 7.19. Path models of motorized/unmotorized boats and type of bank.

Type of bank	Motorized boats and unmotorized boats
Variable X_3 (willow-trees)	The motorized boats show a positive effect with this type of bank in the major part of the areas, viz. in the areas 1, 2, 5, 6, 7, 8, 9, 10, and 11. However, positive values of path coefficients of the unmotorized boats are found in the areas 3, 4, and 11.
Variable X_4 (reed vegetation)	The motorized boats show a positive effect with this type of bank in the areas 1, 2, 4, 7, 9, 10, and 11. The unmotorized boats make use of this type of bank in the areas 3, 5, 6, 7, and 10.
Variable X_5 (landing-stages)	The motorized boats show a positive effect with this type of bank in the areas 1,3,4,5,7,and 8 and the unmotorized boats show a positive effect in the areas 2, 5, 9, 10, and 11.
Variable X_6 (beaches)	A positive effect of the motorized boats on this type of bank is shown in the areas 2, 3, 4, 6,7,10,and 11,and a positive effect of the unmotorized boats is shown in the areas 1, 2, 5, and 8.

The use of motorized boats shows a positive direct effect on the use of willow-trees or reed vegetation in the major part of the areas. The size of the coefficient shows that the strongest direct effects are found with motorized boats.

Having discussed now the stimulus-response relationships with respect to the type of boat (either motorized or unmotorized) and the type of bank, the path coefficients of the recreational activities in terms of type of boat and type of bank will be discussed. Table 7.18(a) shows the path coefficient from Figure 7.3 for each of the 11 Biesbosch areas, and the path coefficients for the other types of bank are shown in Table 7.18 (b)-(d).

The path coefficients of the relationships in Figure 7.3 are discussed in Table 7.20. The coefficients of the recreational activity types X_7 and X_8 (fishing and swimming) are successively discussed in Table 7.20 (a), and the coefficients of the recreational activity types X_9 and X_{10} (activities on the beach) are discussed in Table 7.20 (b).

Table 7.20 (a). Path models of type of boat, type of bank and two types of recreational activities.

Type of bank	Response variable	
	Variable X_7 (e.g. fishing)	Variable X_8 (e.g. swimming)
Variable X_3 (willow-trees)	Motorized boats have a positive effect on the response variable in the areas 1,7,9, 10,and 11. The use of unmotorized boats has a positive effect on the variable X_7 in the areas 2,4, 9,and 10. Variable X_3 also has a positive effect on variable X_7 in the areas 1,2,3,5,6,8,and 11.	The motorized boats show a stronger positive effect on this response variable than with respect to the unmotorized boats. The direct impact of willow-trees on swimming is negative in 7 out of the 11 areas.
Variable X_4 (reed vegetation)	Motorized boats show a positive effect on this variable in the areas 1 and 6, and a negative impact in the areas 2 and 4. The unmotorized boats show a positive effect on variable X_7 in the areas 2,3,4,9,10, and 11. The use of reed vegetation has a positive effect on fishing in the areas 2,3,4,5, 7,8,9,10, and 11.	The use of motorized boats shows a positive effect on variable X_8 in the areas 1, 2,3,4,6,8,and 11. However, the use of unmotorized boats only shows a strong positive effect on variable X_8 in area 2. The use of reed vegetation for swimming has a positive impact in the areas 5,6,7, 9, and 10.
Variable X_5 (landing-stages)	The effect of motorized boats on variable X_7 is in this model larger than with respect to the other types of bank, and is positive in the areas 1,2,5,6,7,9,10, and 11. The effect of unmotorized boats on variable X_7 is positive in the areas 3,4,5,10, and 11. This type of bank only has a positive impact on fishing in the areas 2, and 9.	The use of motorized boats for swimming has a positive effect in most of the areas. The use of unmotorized boats only has a positive effect in the areas 2,9, and 10. The variable X_5 has a negative effect on the variable X_8 in all areas.
Variable X_6 (beaches)	The motorized and unmotorized boats show a positive effect on variable X_7 in the major part of the areas. The use of beaches for fishing only has a positive effect in area 5, and this effect is negative in the major part of the other areas.	The use of motorized boats has a positive effect on this response variable in all areas. The unmotorized boats show a positive direct effect on the variable X_8 in the areas 2 and 10. The use of beaches shows a positive effect on swimming in the areas 2,3,4,5,6,9,11.

Table 7.20 (b). Path models of type of boat, type of bank and two types of recreational activities.

Type of bank	Response variable	
	Variable X_9 (quiet activities)	Variable X_{10} ('unquiet')
Variable X_3 (willow-trees)	The direct effect of motorized boats on the response variable is positive in all areas with the exception of area 9. The direct effect of the use of willow-trees with respect to variable X_9 is negative with the exception of the areas 7,8,9,and 10 in which the effect is positive.	The direct effect of the motorized boats on variable X_{10} is positive in 8 of the 11 areas. The direct effect of the use of willow-trees with respect to variable X_{10} is negative in 8 of the 11 areas. This effect is positive in the areas 5,7, and 9.
Variable X_4 (reed vegetation)	The direct effect of motorized boats on the response variable is larger than the one of unmotorized boats for all areas.	The use of unmotorized boats has a positive direct effect on the response variable in the areas 2,4,5,7 and 8. The variable X_4 has a negative effect on the response variable in all areas with the exception of areas 5 and 10.
Variable X_5 (landing-stages)	The direct effect of motorized boats on the response variable is positive for all areas except area 9. The direct effect of the use of landing-stages on the response variable shows the same pattern as with respect to the response variable X_8, with the exception of area 11 where the direct effect of variable X_5 is positive.	The direct effect of motorized boats on the response variable is only slightly positive in 6 out of the 11 areas. The use of landing-stages has a positive effect on the response variable in the areas 1,3,7,and 11, and this effect parameter is negative in the areas 4,5,8, and 9.
Variable X_6 (beaches)	The direct effect of motorized boats on the response variable is positive for all areas, and the same effect of unmotorized boats is smaller in the 11 areas. The effect of variable X_6 on the 'quiet' activities is positive in the areas 2,4,5, 6,8,and 9, and this effect is negative in the areas 1 and 11.	The direct effect of the variable X_6 on the response variable X_{10} is positive in the areas 2,4,7, and 11

The path models concerning the recreational activities in the Biesbosch area, discussed in Table 7.20, show that the use of motorized boats in almost all areas has a larger direct effect on the recreational activity variables X_8 and X_9 than the one with unmotorized boats. However, the use of unmotorized boats has a larger effect on the variables X_7 and X_{10} than the one with motorized boats for about half of the areas (viz. the areas 2,3,4,5, and 8).

The results discussed in Table 7.20 also show that the four types of bank have different levels of effects with respect to the four types of recreational activities. Consider for example the effect of variable X_6 (the use of beaches by boats) on variable X_8 (activities such as swimming), which is positive in almost all areas, while the direct effect of variable X_5 (the use of landing-stages by boats) on variable X_8 is negative in all areas. This difference in effects indicates that recreational activities such as swimming will primarily be done with beaches as the type of bank.

7.5. CONCLUDING REMARKS

A spatial characterization of outdoor recreation in the Biesbosch area was presented in this chapter. Some of the mathematical tools discussed in Chapter 5 were used. The aspects which are based on outdoor recreation, and presented in the Sections 7.2 to 7.4 are, successively:

(i) a spatial characterization of outdoor recreation in terms of such elements as the area of visit, and the place where the daily purchases are bought (either inside or outside the Biesbosch area). An exploratory and explanatory analysis with log-linear models and linear logit models was presented to define and interpret the statistically significant interactions between variables. These approaches are especially useful for characterizing the regions in terms of significant interaction effects between variables;

(ii) a characterization of phenomena which may be relevant for persons selecting the Biesbosch area for their recreation. A spatial disaggregation was made between five subregions of the Biesbosch, and the phenomena were divided into environmental phenomena, and phenomena which refer to the availability of recreational facilities. The multivariate nature of five phenomena is analysed by means of the HOMALS program;

(iii)a spatial characterization of outdoor recreation patterns with respect to the type of boat used (motor or no motor), the type of riverbank used by the boats (willow-trees, reed-vegetation, landing-stages or beaches), and the specific types of recreational activities (such as fishing or swimming). A multidimensional scaling procedure was used in subsection 7.3.3 to characterize the various subregions of the Biesbosch area. Multivariate analysis has been used in this chapter with respect to the spatial characterization of the Biesbosch area. However, it can also be used in an analogous way to trace temporal trends with respect to multi-variate phenomena.

(iv) an analysis of cause-and-effect relationships in outdoor recreation with respect to type of boat, type of bank and type of recreational activity. This analysis makes use of path models.

Chapter 8. CONCLUSION

8.1. REFLECTION ON THE STUDY

The present study on integrated environmental modelling had the following aims:
- a presentation and discussion of some IEMs;
- development of a model design for an IEM;
- development of mathematical and statistical tools which are relevant to the specific characteristics and problems of operationalising an IEM;
- operationalization of a model design for an integrated environmental approach of the Biesbosch area in the Netherlands.

While achieving these aims, a number of methodological problems were identified. These can be summarized as follows:

(a) complexity in integrating multivariate phenomena which derive from various disciplines;

(b) variables which are measured at a non-metric scale so that statistical procedures based on assumptions such as the normal statistical distribution are inadequate;

(c) insufficient or unreliable data which constrain estimation of parameters in a conventional econometric way.

Such problems were mentioned in the survey of IEMs. However they may also occur in other integrated modelling approaches, and they will therefore be further elaborated in this study.

Some concluding remarks with respect to the aim of the study on the design and the tools of integrated environmental modelling approaches will be presented in this chapter.

An overview of various integrated environmental models, instruments and techniques was presented in part A of the study, viz. in <u>Chapter 2</u> and in <u>Chapter 3</u>, and this survey showed that each of the three main types of environmental categories (viz. water, air and land use) can at least, in principle, be represented by such models. Conclusions drawn from an evaluation of the survey include:

(i) the approach to modelling an IEM depends on the purpose of the study such as description, explanation, forecasting or evaluation;

(ii) the mathematical tools used to operationalize the IEMs tend to vary between optimization, simulation, input/output analysis, scenario analysis, and multi-objective decision analysis;

(iii) the spatial dimension of the model may be regional, or it may be multi-regional;

(iv) limited availability of information may lead to problems in operationalizing one or more of the modules of an IEM. The uncertainty with
respect to information was especially problematical in the case of
modelling environmental phenomena, because of the inadequacy of controlled experiments or of the poor quality of data.

The two basic elements of the study are the design of an IEM and the tools to
operationalize an IEM.

The methodology and tools to operationalize an IEM are presented in Part B,
viz. in Chapter 4 and in Chapter 5, and are based on the discussion of the
IEMs in Part A.

The design of an IEM is presented in Chapter 4, and is based on systems
theory. Systems theory is discussed in this chapter; such integrated models
aim to provide a comprehensive and systematic picture of the components and
interactions between phenomena originating from different sources and disciplines. A systems approach was used because of the a priori hierarchical
consideration of systems and its sub-systems, as well as the input and output
effects between sub-systems. The design of an IEM presented in this chapter
and based on a systems approach was called a satellite design of integration.
The satellite design is characterized by a hierarchical modular representation with one module as the core of the hierarchical analysis, and the
other modules being affected by this core module.

A set of statistical and mathematical tools is discussed in Chapter 5 within
the frame of an IEM.

The first set of tools deal with the analysis of models when only limited information is available on the impacts between variables. Graph theory may be
a useful tool in the case of binary or qualitative information on the stimulus-response relationships between variables. A distinction is made between
the analysis of an hierarchical nature (of systems, when only binary information on variables is available), and analysis of a linear system (of equations when qualitative information with respect to the signs of the parameters is also available). The analysis of binary relationships does not
depend on a normalization rule because it only makes use of information on
whether or not impacts between variables are included. The use of qualitative
calculus becomes especially useful in the case of sign-solvability analysis
from a set of linear equations, which determines the qualitative change of
endogenous variables (in terms of +,-, or 0) with respect to a qualitative
change in the exogenous variables (and lagged endogenous variables). The
conditions of full sign-solvability are rather severe, and additional tools
are presented in Chapter 5 which may lead to at least partial sign-solva-

bility of a set of linear equations. The sign-solvability approach also uses the hierarchical nature of variables as it has been analysed when only binary information is available.

The second set of tools deal with the different sources of data collection, giving a non-metric level of measurement of variables such as the nominal or ordinal level of measurement. A class of statistical models (called generalised linear models) have been developed in the past twenty years to cope with such levels of measurement, and the most widely used ones are log-linear models and linear logit models. Such models can be used for exploratory and explanatory purposes successively.

The third set of tools concerns the multivariate nature of phenomena in the integrated analysis of economic, demographic and environmental phenomena. Such a multivariate nature of phenomena can be represented in terms of nominal or ordinal information. The HOMALS program is useful for nominal information, and the multidimensional scaling procedure is useful for ordinal information. The multivariate approaches are relevant for descriptive or exploratory modelling purposes, depicting spatial patterns or temporal trends.

The design and tools of an IEM discussed in Chapters 4 and 5 are applied in Chapter 6 and Chapter 7 for an integrated modelling approach of the Biesbosch area in the Netherlands.

Outdoor recreation is a key phenomenon in this area, with impacts both on economic activities and environmental phenomena. Land use is the main environmental phenomenon in the analysis (out of the three environmental phenomena water, air and land use). The land use categories included in the analysis are (i) willow-trees and holm, (ii) reed vegetation and saltings, (iii) landing-stages, and (iv) beaches.

A satellite design of integration has therefore been developed with a recreation module, affecting environmental phenomena and economic activities, as the core of the analysis. The advantage of the satellite design of integration is its hierarchical nature, which is in agreement with the systems approach discussed in Chapter 4.

The design of an IEM for the Biesbosch area is discussed in Chapter 6, which is being analysed with different levels of measurement of the impacts between variables. A recreation module is the core module of the satellite design of integration; the regional economy, a natural environment, and demography modules occur as lower level modules.

The main advantages of the satellite design of integration are:

- an a priori selection of relevant phenomena to be analysed, such as the recreational activities in the Biesbosch area as mentioned in this chapter;

- the modular representation of phenomena to give a systematic approach.

The design of an IEM for the Biesbosch area is represented in terms of variables, and this is analysed with binary and qualitative information only. The
model representation in terms of stimulus-response relationships - depicted
in terms of graphs - shows the interdependency between various stages of
variables, with higher level variables determining the stage of variables in
the lower levels. The variables from the regional-economic module (i.e.,
consumption and employment levels with respect to recreational and non-re-
creational activities) are determined by the levels of variables from higher
stages.

A simulation model of an IEM is formulated using qualitative calculus to
determine the qualitative impacts from exogenous variables and lagged endogenous variables on the endogenous variables. This simulation model with 8
equations is not sign-solvable, and additional tools are presented in Chapter
6 by means of a sequential introduction of numerical information on parameter
values. Partial sign-solvability will be achieved with this investigation of
model structure.

The spatial context of outdoor recreation is analysed in Chapter 7. The relevance of various phenomena to come to the Biesbosch area was also discussed
in this chapter, and a distinction is made between the relevance of recreational phenomena, such as the availability of recreational facilities, and
environmental phenomena, such as the presence of plants and animals.

8.2. OUTLOOK OF THE STUDY

The design and mathematical tools are discussed in this study within the
framework of an integrated modelling analysis with respect to the environmental aspects of land use. The satellite design - developed for the Biesbosch area with respect to recreational activities and land use - may also be
useful in the case of an integrated modelling approach for the alternative
environmental phenomena such as water and air. An outline of such a satellite
design concerning an analysis of water distribution and water quality in the
Netherlands was presented in Section 4.4.

The approaches presented in Chapter 5 which are based on the principles of
qualitative calculus - with impacts denoted in qualitative terms - are relevant within the framework of integrated environmental modelling in order to
trace the long-term and/or large-scale environmental consequences of policy
activities and socio-economic development. The main reason for using such
qualitative approaches is the uncertainty with respect to the large scale

and/or long term environmental stimulus-response effects. Consider for example the interactions among ecosystems, socio-economic development and climatic change. Climatic change may affect socio-economic development (such as agricultural and fishery activities). However, these effects can only be covered in a long-term and a large-scale analysis. It is often nearly impossible to obtain insight into the exact changes because of the uncertainties concerning the large temporal and spatial horizons, so that a qualitative approach may be an appropriate vehicle in such cases.

A spatial analysis with respect to the multivariate nature of recreational activities was presented in Chapter 7 by making use of a multidimensional scaling procedure and related statistical techniques for behavioural analysis. Such multivariate analyses can also be used as an exploratory tool to trace spatial patterns and temporal trends in human behaviour regarding environmental phenomena in alternative spatial settings. In conclusion, many methods employed in the present study offer a broad scope for further applications in integrated environmental modelling.

The main achievements of this study are threefold. First, systems theory has shown to be a relevant methodology in case of multidisciplinary environmental research, where systems are described in terms of interdependent stages or processes. It is especially useful in integrated environmental modelling because these models can be represented adequately in terms of interrelated modules with processes which originate from various disciplines. Secondly, the satellite design for integration, which is based on the hierarchical characteristics of systems and subsystems, is a recently developed concept of integration of an IEM. This design is especially useful when one module of an IEM is a key factor of the analysis, and the other modules are related to this core module. Finally, the various mathematical and statistical methods discussed in this study are relevant tools to operationalize an IEM, either with imprecise information concerning parameters or with the various levels of measurement concerning the variables, as well as with the multivariate nature of phenomena in these integrated modelling approaches.

References

Adelman, I. and C. T. Morris, The Derivation of Cardinal Scales from Ordinal Data: an Application of Multidimensional Scaling to Measure Levels of National Development, in: W. Sellekaerts (ed.), Economic Development and Planning, Mac Millan, London, 1974, pp. 1-39.

Aitkin, M., A Simultaneous Test Procedure for Contingency Table Models, Applied Statistics , vol.28, 1979, pp.233-242.

Aitkin, M., A Note on the Selection of Log-linear Models, Biometrics, vol.36, 1980, pp.173-178.

Arntzen, J.W. and L.C. Braat, An Integrated Environmental Model for Regional Policy Analysis, in: T.R. Lakshmanan and P. Nijkamp (eds.), Systems and Models for Energy and Environmental Analysis, Gower, Aldershot, 1983, pp. 45-58.

Arntzen, J.W., L.C. Braat, F. Brouwer and J-P. Hettelingh, Geïntegreerd Milieumodel(Integrated Environmental Model), Institute for Environmental Studies, Free University, Amsterdam, IvM 81/7, 1981 (in Dutch).

Asher, H.B., Causal Modeling, Sage University Paper Series on Quantitative Applications in the Social Sciences, series no.07-003, Beverley Hills and London, Sage Publications, 1983.

Aufhauser, E. and M.M. Fischer, Log-linear Modelling and Spatial Analysis, Environment and Planning A, vol. 17, 1985, pp. 931-951.

Baker, R.J. and J.A. Nelder, The GLIM System: Release 3, Numerical Algorithms Group, Oxford, 1978.

Bartholomew, D.J., Discussion of the paper:Regression Models for Ordinal Data by P. McCullagh, Journal of the Royal Statistical Society Series B, vol.42, 1980, pp.127-129.

Bassett, L., J.S. Maybee and J. Quirk, Qualitative Economics and the Scope of the Correspondence Principle, Econometrica, vol.36, 1968, pp.544-563.

Batey, P.W.J., Information for long-term Planning of Regional Developments, in: P. Nijkamp and P. Rietveld (eds.), Information Systems for Integrated Regional Planning, North-Holland Publ. Company, Amsterdam, 1984, pp. 63-79.

Baumann, J. and U. Schubert, Factors of Regional Labour Force Participation Rates: an Econometric Study for Austria, Paper presented at the Regional Science Association, Munich, 1980.

Beck, M.B., Model Structure Identification from Experimental Data, in: E. Halfon (ed.), Theoretical Systems Ecology: Advances and Case Studies, Academic Press, New York, 1979, pp.260-289.

Beck, M.B., Identifying Models of Environmental Systems' Behaviour, Mathematical Modeling, vol.3, 1982, pp.467-480.

Beguin, H. and J-F. Thisse, An Axiomatic Approach to Geographical Space, Geographical Analysis, vol.11(4), 1979, pp.325-341.

Bennett, R.J. and R.J. Chorley, Environmental Systems: Philosophy, Analysis and Control, Methuen, London, 1978.

Bertalanffy, L. von, General System Theory, in: L. von Bertalanffy and A. Rapoport (eds.), General Systems: Yearbook of the Society for General Systems Research, vol.1, 1956, pp. 1-10.

Bertalanffy, L. von, Robots, Men and Minds: Psychology in the Modern World, George Braziller, New York, 1967.

Bertalanffy, L. von, General System Theory: Foundations, Development, Applications, George Braziller, New York, 1968.

Birch, M.W., Maximum Likelihood in Three-way Contingency Tables, Journal of the Royal Statistical Society Series B, vol.25, 1963, pp.220-233.

Bishop, Y.M., S.E. Fienberg and P.W. Holland, Discrete Multivariate Analysis: Theory and Pratice, MIT Press, Cambridge Massachusetts, 1975.

Blommestein, H. and P. Nijkamp, Multivariate Methods for Soft Data in Development Planning, in: M. Chatterji, P. Nijkamp, T.R. Lakshmanan and C.R. Pathak (eds.), Spatial, Environmental, and Resource Policy in Developing Countries, Gower, Aldershot, 1984, pp. 343-360.

Boulding, K.E., General System Theory - the Skeleton of Science, in : L. von Bertalanffy and A. von Rapoport (eds.), General Systems:Yearbook of the Society for General Systems, vol.1, 1956, pp. 11-17.

Bowlby, S. and J. Silk, Analysis of Qualitative Data using GLIM: two Examples based on Shopping Survey Data, The Professional Geographer, vol. 34 (1), 1982, pp. 80-90.

Boynton, W., D.E. Hawkins and C. Gray, A Modeling Approach to Regional Planning in Franklin County and Apalachicola Bay, Florida, in: C.A.S. Hall and J.W. Day Jr. (eds.), Ecosystem Modeling in Theory and Practice: an Introduction with Case Histories, John Wiley and Sons, New York, 1977, pp.478-505.

Braat, L.C. and W.F.J. van Lierop, A Survey of Economic-Ecological Models, Institute for Environmental Studies - Amsterdam/International Institute for Applied Systems Analysis - Laxenburg, 1984 (mimeographed).

Brouwer, F., J-P. Hettelingh and L. Hordijk, An Integrated Regional Model for Economic-Ecological-Demographic-Facility Interactions, Papers of the Regional Science Association, vol.52, 1983, pp.87-103.

Brouwer, F., W. Hafkamp and P. Nijkamp, Concepts of Integration in Integrated Environmental Models, Man, Environment, Space and Time, vol. 4(2), 1984, pp. 23-54.

Brouwer, F., L. Hordijk and P. Nijkamp, Integrated Regional Economic-Environmental Modeling, in: D.O. Hall, N. Myers and N.S. Margaris (eds.), Economics of Ecosystems Management, Dr. W. Junk Publishers, Dordrecht, 1985, pp.19-29.

Brouwer, F. and P. Nijkamp, Linear Logit Models for Categorical Data in Spatial Mobility Analysis, Economic Geography, vol.60(2), 1984, pp.102-110.

Brouwer, F. and P. Nijkamp, Qualitative Structure Analysis of Complex Systems, in: P. Nijkamp, H. Leitner and N. Wrigley (eds.), Measuring the Unmeasurable, Martinus Nijhoff Publishers, Dordrecht, 1985, pp.509-530.

Brouwer, F. and P. Nijkamp, Model Design in Integrated Regional Environmental Systems, Environments, vol. 18(1), 1986a, pp. 5-13.

Brouwer, F. and P. Nijkamp, A Satellite Design for Integrated Regional Environmental Modelling, Ecological Modelling, 1986b (forthcoming).

Brouwer, F. and P. Nijkamp, Sign-solvability Analysis with Qualitative and Quantitative Information, Kwantitatieve Methoden,vol. 20,1986c, pp.37-47.

Brouwer, F. and P. Nijkamp, A Satellite Design for Integrated Land Use and Water Resource Management Models, in: L. Valadares Tavares (ed.), Systems Analysis Applied to Water and Related Land Resources, Pergamon Press, Oxford, 1986d (forthcoming).

Brouwer, F. and P. Nijkamp, Mixed Qualitative Calculus as a Tool in Policy Modeling: a Dynamic Simulation Model of Urban Decline, Journal of Policy Modeling, vol. 8 (1),1986e, pp. 69-88.

Brown, M., War, Peace and the Computer: Simulation of Disordering and Ordering Energies in South Vietnam, in: C.A.S. Hall and J.W. Day Jr. (eds), Ecosystems Modelling in Theory and Practice: an Introduction with Case Histories, John Wiley and Sons, New York, 1977, pp.394-417.

Brown, M.B., Screening Effects in Multidimensional Contingency Tables, Applied Statistics, vol.25, 1976, pp.37-46.

Brown, M.B., BMDP routine P4F: Two-way and Multiway Frequency Tables - Measures of Association and the Log-linear Model (Complete and Incomplete Tables), in: W.J. Dixon, (ed.), BMDP Statistical Software 1981, University of California Press, Los Angeles, 1981.

Burch, J.G., F.R. Strater and G. Grudnitski, Information Systems: Theory and Practice, John Wiley, New York, 1979.

Cadwallader, M., Structural Equation Models in Human Geography, Progress in Human Geography, vol. 10(1), 1986, pp. 24-47.

Casti, J.L., Connectivity, Complexity and Catastrophy in Large-Scale Systems, John Wiley and Sons, New York, 1979.

Casti, J.L., On the Theory of Models and the Modelling of Natural Phenomena, in: G. Bahrenberg, M.M. Fischer and P. Nijkamp (eds.), Recent Developments in Spatial Data Analysis: Methodology, Measurement, Models, Gower Publ. Comp., Aldershot, 1984, pp.73-84.

Caswell, H., H.E. Koenig, J.A. Resh and Q.E. Ross, An Introduction to Systems Science for Ecologists, in: B.C. Patten (ed.), Systems Analysis and Simulation in Ecology, vol.2, Academic Press, New York, 1972, pp.3-78.

Clark, W.C., Sustainable Development of the Biosphere: Themes for a Research Program, in: W.C. Clark and R.E. Munn (eds.), Sustainable Development of the Biosphere, Cambridge University Press, Cambridge, 1986, pp. 5-48.

Clark, W.C., Scales of Climate Impacts, Climatic Change, vol. 7, 1985, pp. 5-27.

Coolen, H.C.C.H. and J.L.A. van Rijckevorsel, Dutch Law and Order Mentality in the Seventies, Erasmus University, Faculty of Social Sciences, Dept. AOA, AOA/1981/03, Rotterdam, 1981.

Coombes, M.G., J.S. Dixon, J.B. Goddard, S. Openshaw and P.J. Taylor, Functional Regions for the Population Census of Great Brittain, in: D.T. Herbert and R.J. Johnston (eds.), Geography and the Urban Environment, Progress in Research and Applications (vol.5), John Wiley and Sons, New York, 1982, pp.63-112.

Courbis, R., The REGINA Model: a Regional-National Model for French Planning, Regional Science and Urban Economics, vol. 9, 1979, pp.117-139.

Curry, L., Elements of Spatial Statistical Systems Analysis, Professional Geographer, vol.35, 1983, pp.149-157.

Duckstein, L., I. Bogardi and L. David, Multiobjective Control of Nutrient Loading into a Lake, in: Y. Haimes and J. Kindler (eds.), Water and Related Land Resource Systems, Pergamon Press, Oxford, 1980, pp.413-418.

Duckstein, L., I. Bogardi and L. David, Dual Objective Control of Nutrient Loading into a Lake, Water Resources Bulletin, vol.18 (1), 1982, pp.21-26.

Duncan, O.D., Path Analysis: Sociological Examples, in: H.M. Blalock (ed.), Causal Models in the Social Sciences, Aldine Publishing Company, Chicago, 1971, pp.115-138.

Eriksson, K-E., Thermodynamical Aspects on Ecology/Economics, in: A-M. Jansson (ed.), Integration of Economy and Ecology: an Outlook for the Eighties, Proceedings from the Wallenberg Symposia, Stockholm, 1984, pp.39-45.

Everitt, B.S., The Analysis of Contingency Tables, Chapman and Hall, London, 1977.

Fay, R.E. and L.A. Goodman, The ECTA Program: Description for Users, Department of Statistics, University of Chicago, 1975.

Fedra, K., Environmental Modeling under Uncertainty: Monte Carlo Simulation, International Institute for Applied Systems Analysis, RR-83-28, Laxenburg, 1983.

Fedra, K., G. van Straten and M.B. Beck, Uncertainty and Arbitrariness in Ecosystems Modelling: a Lake Modelling Example, Ecological Modelling, vol.13, 1981, pp.87-110.

Fienberg, S.E., An Iterative Procedure for Estimation in Contingency Tables, Annals Mathematical Statistics, vol.41, 1971, pp.907-917.

Fienberg, S.E., The Analysis of Cross-Classified Categorical Data, 2nd edition, MIT Press, Cambridge Massachusetts, 1980.

Fienberg, S.E. and M.M. Meyer, Log-linear Models and Categorical Data Analysis with Psychometric and Econometric Applications, Journal of Econometrics, vol.22, 1983, pp.191-214.

Fingleton, B., Log-linear Modelling of Geographical Contingency Tables, Environment and Planning A, vol.13, 1981, pp.1539-1551.

Fischer, M.M., Zur Lösung Funktionaler Regionaltaxonomischer Probleme auf der Basis vor Interaktionsmatrizen: ein neuer Graphentheoretischer Ansatz, Karlsruher Manuskripte zur Geographie, Geographisches Institut, heft 25, Karlsruhe, 1978.

Fischer, M.M., Regional Taxonomy: a Comparison of some Hierarchic and Non-Hierarchic Strategies, Regional Science and Urban Economics, vol. 10, 1980, pp. 503-537.

Fischer, M.M., Regional Taxonomy: some Reflections on the State of the Art, paper presented at the International Seminar on 'Regional Planning and Information System Requirements in Regional Planning', Cagliari, 1983.

Fischer, M.M. and P. Nijkamp, Developments in Explanatory Discrete Spatial Data and Choice Analysis, Progress in Human Geography vol. 9, 1985, pp. 515-551.

Fisher, A.C. and F.M. Peterson, The Environment in Economics: A Survey, Journal of Economic Literature, vol. 14, 1976, pp.1-33.

Forrester, J.W., World Dynamics, Wright Allen,Cambridge Massachusetts, 1971.

Gale, N. and R.G. Golledge, On the Subjective Partitioning of Space, Annals of the Association of American Geographers, vol.72 (1), 1982, pp.60-67.

Ganin, I.A. and D.P. Solomatin, STRUM - An Interactive Computer System for Modeling Binary Relations, Collaborative Paper CP-84-52, International Institute for Applied Systems Analysis, Laxenburg, Austria, 1984.

Garbely, M. and M. Gilli, Two Approaches in Reading Model Interdependencies, in: J.P. Ancot (ed.), Analysing the Structure of Econometric Models, Series Advanced Studies in Theoretical and Applied Econometrics, vol.2, Martinus Nijhoff Publishers, The Hague, 1984, pp.15-33.

Gifi, A., HOMALS Users Guide, University of Leiden, Faculty of Social Sciences, Department of Datatheory, Leiden, 1981.

Gilchrist, R., (ed.), GLIM 82: Proceedings of the International Conference on Generalised Linear Models, Lecture Notes in Statistics, vol.14, Springer-Verlag, New York/Berlin, 1982.

Gilli, M., CAUSOR, a Program for the Analysis of Recursive and Interdependent Causal Structures, Cahier 84.03, Department of Econometrics, University of Geneva, 1984.

Gilli, M. and E. Rossier, Understanding Complex Systems, Automatica, vol.17 (4), 1981, pp.647-652.

Goeller, B.F. et al., Projecting an Estuary from Floods - a Policy Analysis of the Oosterschelde (vol.1, Summary Report), R-2121/1-Neth, Rand Corporation, Santa Monica, 1977.

Goeller, B.F. et al., Policy Analysis of Water Management for the Netherlands (vol.1, Summary Report), R-2500/1-Neth, Rand Corporation, Santa Monica, 1983.

Goodall, D.W., The Hierarchical Approach to Model Building, in: G.W. Arnold and C.T. de Wit (eds.), Critical Evaluation of Systems Analysis in Ecosystems Research and Management, Centre for Agricultural Publishing and Documentation, Wageningen, 1976, pp.10-21.

Goodman, L.A., The Analysis of Multidimensional Tables: Stepwise Procedures and Direct Estimation Methods for Building Models for Multiple Classifications, Technometrics, vol.13, 1971, pp.33-61.

Goodman L. A., The Analysis of Multidimensional Contingency Tables when some Variables are posterior to others: a Modified Path Analysis Approach, Biometrika vol. 60(1), 1973, pp. 179-192.

Gottinger, H.W. (ed.), Systems Approaches and Environmental Problems, Vandenhoeck and Ruprecht, Göttingen, 1974.

Greenberg, H.J. and J.S. Maybee (eds.), Computer-Assisted Analysis and Model Simplification, Academic Press, New York, 1981.

Grizzle, J.E., C.F. Starmer and G.G. Koch, Analysis of Categorical Data by Linear Models, Biometrics, vol.25, 1969, pp.489-504.

Hafkamp, W.A., Economic-Environmental Modeling in a National-Regional System, North-Holland Publishing Company-Amsterdam, 1984.

Hagenaars, J.A.P., Loglineaire Analyse van Herhaalde Surveys: Panel-, Trend-en Cohortononderzoek, Department of Social Science, University of Til-burg (Ph.D. Thesis), 1985 (in Dutch).

Hall, G.B. and S.M. Taylor, A Causal Model of Attitudes toward Mental Health Facilities, Environment and Planning A, vol.15(4), 1983, pp.525-542.

Hordijk, L., H.M.A. Jansen, A.A. Olsthoorn and J.B.Vos, Reken- en Informatie-systeem Milieuhygiëne - een studie naar haalbaarheid, Ministry of Public Health and Environment, Series: VAR 1981/13, State Publishing Company, The Hague, 1981 (in Dutch).

Ikeda, S., Economic-ecological Models in Regional Setting, Institute of So-cio-Economic Planning, University of Tsukuba, Sakura, Discussion Paper, no.241, 1984.

Ikeda, S. and H. Itakura, Multidimensional Scaling Approach to Clustering Multivariate Data for Water-quality Modeling, in: M.B. Beck and G. van Straten (eds.), Uncertainty and Forecasting of Water Quality, Springer-Verlag, Berlin/Heidelberg, 1983, pp.205-223.

Imrey, P.B., G.G. Koch and M.E. Stokes, Categorical Data Analysis: Some Re-flections on the Log-linear Model and Logistic Regression. Part I: His-torical and Methodological Overview, International Statistical Review, vol.49, 1981, pp.265-283.

Imrey, P.B., G.G. Koch and M.E. Stokes, Categorical Data Analysis: Some Re-flections on the Log-linear Model and Logistic Regression, Part II: Data Analysis, International Statistical Review, vol.50, 1982, pp. 35-63.

Isard, W., On the linkage of Socio-economic and Ecologic Systems, Papers of the Regional Science Association, vol. 21, 1968, pp. 79-99.

Israëls, A.Z., CORAN-HOMALS-CANALS, Netherlands Central Bureau of Statistics, Voorburg, 1981 (mimeographed).

Israëls, A.Z., Path Analysis for Mixed Qualitative and Quantitative Varia-bles, Kwantitatieve Methoden, vol. 6 (4), 1985, pp. 73- 87.

Issaev, B., P. Nijkamp, P. Rietveld and F. Snickars (eds.), Multiregional Economic Modelling: Practice and Prospect, North-Holland Publishing Company, Amsterdam, 1982.

Jansson, A-M. (ed.), Integration of Economy and Ecology: an Outlook for the Eighties, Proceedings from the Wallenberg Symposia, Stockholm, 1984.

Jeffers, J.N.R., Future Prospects of Systems Analysis in Ecology, in: G.W. Arnold and C.T. de Wit (eds.), Critical Evaluation of Systems Analysis in Ecosystems Research and Management, Centre for Agricultural Publish-ing and Documentation, Wageningen, 1976, pp.98-108.

Jeffers, J.N.R., Modelling, Series: Outline Studies in Ecology, Chapman and Hall, London, 1982.

Jeffries, C., Qualitative Stability and Digraphs in Model Ecosystems, Ecology, vol.55, 1974, pp.1415-1419.

Jones, G.E., The Usefulness of System Theory in Ecosystem Studies, Area, vol.15(2), 1983, pp.111-116.

Jørgensen, S.E., State of the Art of Ecological Modelling, Sistemi Urbani, vol. 5 (1), 1983, pp.107-117.

Kelly, R.A. and W.O. Spofford Jr, Application of an Ecosystem Model to Water Quality Management: the Delaware Estuary, in: C.A.S. Hall and J.W Day Jr. (eds.), Ecosystem Modeling in Theory and Practive: an Introduction with Case Histories, John Wiley and Sons, New York, 1977, pp.420-443.

Kendall, M.G. and G.U. Yule, An Introduction to the Theory of Statistics, Charles Griffin, London, 1950.

Klein, L.R., The Specification of Regional Econometric Models, Papers of the Regional Science Association, vol. 23, 1969, pp.105-115.

Klein, L.R. and N.J. Glickman, Econometric Model-building at Regional Level, Regional Science and Urban Economics, vol.7, 1977, pp.3-23.

Kneese, A.V., R.U. Ayres and R.C. D'Arge, Economics and the Environment: a Materials Balance Approach, Resources for the Future, The Johns Hopkins University Press, Baltimore/London, 1970.

Kruskal, J.B. and M. Wish, Multidimensional Scaling, Sage University Paper series on Quantitative Applications in the Social Sciences, 07-011. Beverley Hills and London, Sage Publications, 1978.

Lady, G.M., The Structure of Qualitatively Determinate Relationships, Econometrica, vol.51, 1983, pp.197-218.

Lakshmanan, T.R., A Multiregional Model of the Economy, Environment, and Energy Demand in the United States, Economic Geography, vol. 59, 1983, pp.296-320.

Lakshmanan, T.R. and R. Bolton, Regional Energy and Environmental Analysis, in: P. Nijkamp and E. Mills (eds.), Handbook in Regional and Urban Economics, vol. 1: Regional Economics, North-Holland Publishing Company , Amsterdam, 1986 (forthcoming).

Lakshmanan, T.R. and S. Ratick, Integrated Models for Economic-Energy-Environmental Impact Analysis, in: T.R. Lakshmanan and P. Nijkamp (eds.), Economic-Environmental-Energy Interactions, Martinus Nijhoff, Boston, 1980, pp.7-39.

Lammerts van Bueren, W.M., Measuring Association in Nominal Data, Department of Economics, Erasmus University, Rotterdam (Ph.D. Thesis), 1982.

Lancaster, K., The Scope of Qualitative Economics, Review of Economic Studies, vol.29, 1962, pp.99-123.

Landis, J.R., W.M. Stanish, J.L. Freeman and G.G. Koch, A Computer Program for the Generalized Chi-square Analysis of Categorical Data Using Weighted Least .Squares (GENCAT), Computer Programs in Biomedicine, vol. 6, 1976, pp.196-231.

Langbeim, L.I. and A.J. Lichtman, Ecological Inference, Sage University Paper Series on Quantitative Applications in the Social Sciences, 07-010, Beverley Hills and London, Sage Publications, 1978.

Légrády, K., The PAWN Models of the Netherlands Rijkswaterstaat; a Case Study of the Methodological Aspects, Report 83-23, Delft University of Technology, Department of Mathematics and Informatics, Delft, 1983.

Leitner, H. and H. Wohlschlägl, Metrische und Ordinale Pfadanalyse: ein Verfahren zur Testung Komplexer Kausalmodelle in der Geographie, Geographische Zeitschrift, vol.68(2), 1980, pp.81-106.

Leitner, H., E. Sheppard and H. Wohlschlägl, Generalized Path Analysis for Mixed Geographical Data, in: P. Nijkamp, H. Leitner and N. Wrigley (eds.), Measuring the Unmeasurable, Martinus Nijhoff, Dordrecht, 1985, pp. 371-397.

Lesuis, P., F. Muller and P. Nijkamp, An Interregional Policy Model for Energy-Economic-Environmental Interactions, Regional Science and Urban Economics, vol. 10, 1980, pp. 343-370.

Lierop, W.F.J. van, Spatial Interaction Modelling and Residential Choice Analysis, Gower, Aldershot, 1986.

Linden, J.W. van der and T. van Eijk, Recreatiegedrag en recreatiepatroon, (Recreational Behaviour and Recreational Pattern), Institute for Environmental Studies, Free University, Amsterdam, 1984 (in Dutch).

Lohmoeller, J-B., J.W. Falter, A. Link and J. de Rijke, Unemployment and the Rise of National Socialism: Contradicting Results for Different Regional Aggregations, in: P. Nijkamp, H. Leitner and N. Wrigley (eds.), Measuring the Unmeasurable, Martinus Nijhoff Publishers, Dordrecht, 1985, pp.357-370.

Lonergan, S.C., A Methodological Framework for Resolving Ecological/Economic Problems, Papers of the Regional Science Association, vol.48, 1981, pp.117-133.

Lonergan, S.C., A Simulation/Optimization Model for Natural Resource Planning: the Chesapeake Bay Experience, Resource Management and Optimization, vol. 2 (4), 1983, pp.293-321.

Lundqvist, L., Applications of a Dynamic Multiregional Input-Output Model of the Swedish Economy, Papers of the Regional Science Association, vol. 47, 1981, pp.77-95.

Macdonald, K.I.,Path Analysis, in: C.A. O'Muircheartaigh and C. Payne (eds.), The Analysis of Survey Data: Model Fitting (vol.2), John Wiley and Sons, Chichester/New York, 1977, pp.81-104.

Man, W.H. de (ed.), Conceptual Framework and Guidelines for Establishing Geographic Information Systems Capable of Integrating Natural Resources Data and Socio-economic Data for Development-oriented Planning, Monitoring and Research, General Information Programme and UNISIST, UNESCO, Paris, 1984.

Man and Biosphere (MAB), Interactions between Ecological, Economical and Social Systems in Regions of Intensive Agriculture, MAB-vol.7 Bonn, 1982.

Man and Biosphere (MAB), Modelling of the Socio-economical and Ecological Consequences of High Animal Waste Application, MAB-vol.14, Bonn, 1983.

Maybee, J.S., Sign-Solvability, in: H.J. Greenberg and J.S. Maybee (eds.)., Computer-Assisted Analysis and Model Simplification, Academic Press, New York, 1981, pp.201-257.

Maybee, J.S. and H. Voogd, Qualitative Impact Analysis through Sign-solvability: a Review, Environment and Planning B: Planning and Design, vol.11, 1984, pp.365-376.

McConnell, K.E., The Economics of Outdoor Recreation, in : A.V. Kneese and J.L. Sweeney (eds.), Handbook of Natural Resource and Energy Economics, vol. 2, North-Holland Publishing Company, Amsterdam, 1985, pp. 677-722.

McCullagh, P. and J.A. Nelder, Generalized Linear Models, Chapman and Hall, London, 1983.

Mesarović, M.D. (ed.), Views on General Systems Theory, Proceedings of the second systems symposium at Case Institute of Technology, John Wiley and Sons, New York, 1964.

Molenaar, I.W., Statistics in the Social and Behavioral Sciences, Statistica Neerlandica, vol. 39 (2), 1985, pp. 169-179.

Morgan, R.K., Systems Analysis: a Problem of Methodology?, Area, vol.13(3), 1981, pp.219-230.

Müller, N., Hierarchical-sequential Decomposition: a Comprehensive Approach for Real-Structure Modelling of Social Systems, in: Cellier (ed.), Progress in Modelling and Simulation, Academic Press,New York,1981, pp.85-101.

Müller, N., Real Structure Modelling: towards a Valid Approach for Social Systems Analysis, in: H. Wedde (ed.), Adequate Modeling of Systems, Springer-Verlag, Berlin, 1983, pp.279-295.

Müller, N., Modelling Standard Actions of Individuals and Institutions for Controlling NO_3-Concentration in Drinking Water in a Region of Intensive Agriculture, University of Osnabrück, 1985 (mimeographed).

Murphy, P. E., _Tourism: a Community Approach_, Methuen, London, 1985.

Nadler, G., Systems Methodology and Design, _IEEE Transactions on Systems, Man, and Cybernetics_, vol. 15(6), 1985, pp. 685-697.

Nelder, J.A., Log-linear Models for Contingency Tables: a Generalization of Classical Least Squares, _Applied Statistics_, vol.23, 1974, pp.323-329.

Nelder, J.A., Statistical Models for Qualitative Data, in: P. Nijkamp, H. Leitner and N. Wrigley (eds.), _Measuring the Unmeasurable_, Martinus Nijhoff Publishers, Dordrecht, 1985, pp. 31-38.

Nelder, J.A. and R.W.M. Wedderburn. Generalized Linear Models, _Journal of the Royal Statistical Society_, Series A, vol.135, 1972, pp.370-384.

Nijkamp, P., _Theory and Application of Environmental Economics_, North-Holland Publishing Company, Amsterdam, 1977.

Nijkamp, P., Soft Econometric Models: an Analysis of Regional Income Determinants, _Regional Studies_, vol.16, 1982, pp.121-128.

Nijkamp, P., Information Systems for Regional Development Planning: a State-of-the Art Survey, _Environment and Planning, B: Planning and Design_, vol. 10, 1983, pp. 283-302.

Nijkamp, P., Economic and Ecological Models: a Qualitative Multidimensional View, in: A-M. Jansson (ed.), _Integration of Economy and Ecology: an Outlook for the Eighties_, Proceedings from the Wallenberg Symposia, Stockholm, 1984a, pp. 167-184.

Nijkamp, P., Information Systems: a General Introduction, in: P. Nijkamp and P. Rietveld (eds.), _Information Systems for Integrated Regional Planning_, North-Holland Publishing Company, Amsterdam, 1984b, pp. 3-33.

Nijkamp, P., H. Leitner and N. Wrigley (eds.), _Measuring the Unmeasurable_, Martinus Nijhoff Publishers, Dordrecht, 1985.

Nijkamp, P. and J.B. Opschoor, Naar een geïntegreerd milieumodel? (Towards an Integrated Environmental Model?), Working Paper IvM-A18, Institute for Environmental Studies, Free University, Amsterdam, 1977 (in Dutch).

Nijkamp, P. and P. Rietveld, Structure Analysis of Spatial Systems, in: B. Issaev, P. Nijkamp, P. Rietveld and F. Snickars (eds.), _Multiregional Economic Modeling: Practice and Prospects_, North-Holland Publishing Company, Amsterdam, 1982, pp.35-48.

Nijkamp, P., P. Rietveld and A. Rima, Information Content of Data from Different Spatial Aggregation Levels, in: P. Nijkamp and P. Rietveld (eds.), _Information Systems for Integrated Regional Planning_, North-Holland Publishing Company, Amsterdam, 1984, pp.215-228.

Nijkamp, P., P. Rietveld and F. Snickars, The Use of Regional and Multiregional Economic Models, in: P. Nijkamp and E. Mills (eds.), _Handbook of Regional and Urban Economics_, vol. 1: Regional Economics, North-Holland Publishing Company, Amsterdam, 1986 (forthcoming).

Nijkamp, P. and H. Voogd, Multidimensional and Homogeneous Scaling in Spatial Analysis, in: G. Bahrenberg, M.M. Fischer and P. Nijkamp, Recent Developments in Spatial Data Analysis: Methodology, Measurement, Models, Gower Publ. Comp., Aldershot, 1984, pp.201-214.

Nishikawa, Y., S. Ikeda, N. Adachi, A. Udo and H. Yukawa, An Ecologic-economic Model for Supporting Land-marine Integrated Development - in the Case of the East Seto Inland Sea, in: Y. Haimes and J. Kindler (eds.), Water and Related Land Resource Systems, Pergamon Press, Oxford, 1980, pp.141-149.

Nishikawa, Y., Sannomiya N., M. Maeda and T. Kitamura, Multi-objective Linear Programming Model as Applied to Regional Land-Use Planning, in: T. Hasegawa (ed.)., IFAC Workshop on Urban Regional and National Planning (UN-RENAP); Environmental Aspects, Pergamon Press, Oxford,1977, pp.161-168.

Norgaard, R., Environmental Economics: An Evolutionary Critique and a Plea for Pluralism, Journal of Environmental Economics and Management, vol. 12, 1985, pp. 382-394.

O'Brien, L.G. and N. Wrigley, A Generalized Linear Models Approach to Categorical Data Analysis: Theory and Applications in Geography and Regional Science, in: G. Bahrenberg, M.M. Fischer and P. Nijkamp, Recent Developments in Spatial Data Analysis: Methodology, Measurement, Models, Gower Publ. Comp., Aldershot, 1984, pp.231-251.

Odum, H.T., Environment, Power and Society, John Wiley, New York, 1971.

Openshaw, S., The Modifiable Areal Unit Problem, Series: Concepts and Techniques in Modern Geography, vol.38, Geo Books, Norwich, 1983.

Openshaw, S. and P.J. Taylor, A Million or so Correlation Coefficients: Three Experiments on the Modifiable Areal Unit Problem, in: N. Wrigley (ed.), Statistical Applications in the Spatial Sciences, Pion, London, 1979, pp.127-144.

Openshaw, S. and P.J. Taylor, The Modifiable Areal Unit Problem, in: N. Wrigley and R.J. Bennett (eds.), Quantitative Geography: a British View, Routledge and Kegan Paul, London, 1981, pp.60-69.

Paelinck, J.H. and P. Nijkamp, Operational Theory and Method in Regional Economics, Saxon House, Farnborough, 1976.

Pannekoek, J. and L. H. Stronkhorst, Socio- economic Status and School Career: Explorations with Log-linear Models, Netherlands Bureau of Statistics,Department for Statistical Methods, Voorburg, 1981 (mimeographed).

Payne, C., The Log-linear Model for Contingency Tables, in: C. Payne and C.A. O'Muircheartaigh (eds.), The Analysis of Survey Data, vol.2: Model Fitting, John Wiley and Sons, New York, 1977, pp.109-144.

Pindyck, R.S. and D.L. Rubinfeld, Econometric Models and Economic Forecasts, McGraw-Hill, New York, 1976.

Ploeg, S.W.F. van der, L.C. Braat, W.F.J. van Lierop and J. van der Linden, Openluchtrecreatie en natuurlijk milieu in de Biesbosch, (Outdoor recreation and the natural environment in the Biesbosch area), Institute for Environmental Studies, Free University, Amsterdam, 1984 (in Dutch).

Rahmatian, S., The Hierarchy of Objectives: Toward an Integrating Construct in Systems Science, Systems Research, vol. 2(3), 1985, pp. 237-245.

Richardson, H.W., The State of Regional Economics: A Survey Article, International Regional Science Review, vol. 3(1), 1978, pp. 1-48.

Rietveld, P., Information Systems for Regional Labor Markets, in : P. Nijkamp and P. Rietveld (eds.), Information Systems for Integrated Regional Planning, North-Holland Publishing Company, Amsterdam, 1984, pp. 163-177.

Ritschard, G., ANAS - A Program for Analysing and Solving Qualitative Systems, Department of Econometrics, University of Geneva, Geneva, 1980.

Ritschard, G., Computable Qualitative Comparative Static Techniques, Econometrica, vol.51, 1983, pp.1145-1168.

Roberts, F.S., Graph Theory and its Applications to Problems of Society, Series: Society for Industrial and Applied Mathematics, vol.29, Philadelphia, Pennsylvania, 1978.

Rosen, R., Some Systems Theoretical Problems in Biology, in: E. Laszlo (ed.), The Relevance of General Systems Theory: Papers Presented to Ludwig von Bertalanffy on his Seventieth Birthday, George Braziller, New York, 1972, pp. 43-66.

Royer, D. and G. Ritschard, Qualitative Structural Analysis: Game or Science?, in: J.P. Ancot (ed.), Analysing the Structure of Econometric Models, Series: Advanced Studies in Theoretical and Applied Econometrics, vol.2, Martinus Nijhoff Publishers, The Hague, 1984, pp.3-13.

Russell, C. S. and W.O. Spofford Jr., A Regional Environmental Quality Management Model: an Assessment, Journal of Environmental Economics and Management , vol.4, 1977, pp.89-110.

Samuelson, P., Foundations of Economic Analysis, Harvard University Press, Cambridge, 1947.

Schmidt, P. and R.P. Strauss, The Prediction of Occupation Using Multiple Logit Models, International Economic Review, vol.16, 1975, pp.471-486.

Siebert, H., Spatial Aspects of Environmental Economics, in: A.V. Kneese and J.L. Sweeney (eds.), Handbook of Natural Resource and Energy Economics, vol. 1, North-Holland Publishing Company, Amsterdam, 1985, pp. 125-164.

Simon, H.A., The Organization of Complex Systems, in: H.H. Pattee (ed.), Hierarchy Theory, Braziller, New York, 1973, pp. 3-27.

Solomon, B.D., Regional Econometric Models for Environmental Impact Assessment, Progress in Human Geography, vol. 10 (3), 1985, pp. 379-399.

Solomon, B.D. and B.M. Rubin, Environmental Linkages in Regional Econometric Models: an Anlysis of Coal Development in Western Kentucky, Land Economics, vol. 61 (1), 1985, pp. 43-57.

Somermeyer, W.H., Specificatie van Economische Relaties (Specification of Economic Relationships), De Economist, vol. 115 (3), 1967, pp. 305-327.

Spiller, G., A-M. Jansson and J. Zucchetto, Modelling the Effects of Regional Energy Development on Groundwater Nitrate Pollution on Gotland, Sweden, in: W.J. Mitsch, R.W. Basserman and J.M. Klopatek (eds.), Energy and Ecological Modelling, Series: Developments in Environmental Modelling, vol.1, Elsevier Scientific Publishing Company, Amsterdam, 1981, pp.495-506.

Spofford, W.O., C.S. Russell and R.A. Kelly, Environmental Quality Management: an Application to the Lower Delaware Valley, Resources for the Future, Washington D.C., 1976.

Steinitz, C., H.J. Brown, P. Goodale and P. Rogers, Managing Suburban Growth: a Modeling Approach, Landscape Architecture Research Office, Graduate School of Design, Harvard University, Cambridge, Massachusetts, 1976.

Steinitz, C. and H.J. Brown, A Computer Modeling Approach to Managing Urban Expansion, Geo-Processing, vol.1, 1981, pp.341-375.

Svedin, U., Economic and Ecological Theory: Differences and Similarities, in: D.O. Hall, N. Myers and N.S. Margaris (eds.), Economics of Ecosystems Management, Dr. W. Junk Publishers, Dordrecht, 1985, pp.31-39.

Tai, K.C., Flood Plain Management Models for Economic, Environmental and Ecological Impact Analysis, in: S.E. Jørgensen (ed.), State-of-the-Art in Ecological Modelling, vol.7, International Society for Ecological Modelling, 1979, pp.405-417.

Takeuchi, K., H. Yanai and B.N. Mukherjee, The Foundations of Multivariate Analysis: a Unified Approach by Means of Projection onto Linear Subspaces, Wiley Eastern Limited, New Delhi, 1982.

Theil, H., On the Estimation of Relationships Involving Qualitative Variables, American Journal of Sociology, vol.76, 1970, pp.103-154.

Tinkler, K.J., Introduction to Graph Theoretical Methods in Geography, Series: Concepts and Techniques in Modern Geography, vol.14, Institute of British Geographers, London, 1977.

Wermuth, N. and S. L. Lauritzen, Graphical and Recursive Models for Contingency Tables, Report R 82-2, Institute for Elektroniske Systemer, Aalborg Universitetscenter, Aalborg, Danmark, 1982.

Whittam, T.S. and D. Siegel-Causey, Species Interactions and Community Structure in Alaskan Seabird Colonies, Ecology, vol.62(6), 1981, pp.1515-1524.

Wilson, A.G., Geography and the Environment: Systems Analytical Methods, John Wiley and Sons, Chichester/New York, 1981.

Wrigley, N., Developments in the Statistical Analysis of Categorical Data, Progress in Human Geography, vol.3, 1979, pp.315-355.

Wrigley, N., Categorical Data, Repeated-Measurement Research Designs and Regional Industrial Surveys, Regional Studies, vol.14, 1980, pp. 455-471.

Wrigley, N., Categorical Data Analysis for Geographers and Environmental Scientists, Longman, London/New York, 1985.

Wrigley, N. and F. Brouwer, Qualitative Statistical Models for Regional Economic Analysis, in: P. Nijkamp and E. Mills (eds.), Handbook of Regional and Urban Economics, vol.1: Regional Economics, North-Holland Publ. Comp., Amsterdam, 1986 (in press).

Wymore, A.W., Systems Engineering Methodology for Interdisciplinary Teams, John Wiley and Sons, New York, 1976.

Zucchetto, J. and A-M. Jansson, Total Energy Analysis of Gotland's Agriculture: a Northern Temperate Zone Case Study, Agro-Ecosystems, vol.5, 1979, pp.329-344.

Zucchetto, J. and A-M. Jansson, Systems Analysis of the Present and Future Energy/Economic Development on the Island of Gotland, Sweden, in: W.J. Mitsch, R.W. Basserman and J.M. Klopatek (eds.), Energy and Ecological Modelling, Series: Developments in Environmental Modelling, vol.1, Elsevier Scientific Publishing Company, Amsterdam, 1981, pp.487-494.

AUTHOR INDEX

SUBJECT INDEX